The Sociology of Developing Societies

The Sociology of Developing Societies

Ankie M. M. Hoogvelt

Lecturer in Sociology, Department of Sociological Studies, University of Sheffield

M

First published 1976 by
THE MACMILLAN PRESS LTD
London and Basingstoke
Associated companies in New York Dublin
Melbourne Johannesburg and Madras

SBN 333 15758 3 (hard cover)
333 15759 1 (paper cover)

Printed in Great Britain by
WESTERN PRINTING SERVICES LTD
Bristol

Voor Pappa

Contents

Preface

It is a good convention to use a preface to pay tribute to others, and I am often impressed and a little surprised to read the ease with which other authors seem able to trace the origins and the personal history of their ideas to specific friends and colleagues.

My own situation is different for two reasons. First, there is very little that is new or original in this book, as the text is patently designed to be an introductory guide to ideas and theories which have been around for some time. The sources of these ideas and theories, where not explicitly mentioned in the text, can be found in the reference section at the end of the book. I have merely strived to bring these ideas together and to present them in a concise and systematic form which I felt to be lacking in the field. Yet, I owe the ability and the confidence to do just that to those who taught me – especially the scholarly folk at the Institute of Social Studies in The Hague – as well as to those around me who create an encouraging and intellectually stimulating environment: my colleagues, students and friends at the University of Sheffield. Both groups of people are too numerous to mention individually, and I would not like to miss anyone out. But I am grateful to them all.

On the other hand, the compelling need to write on the subject of development, and my understanding of the concrete realities which sometimes defy, sometimes painfully fit, the development theories, I *can* trace quite clearly to just such a handful of friends: friends in the developing countries themselves. Caught in a web of contradictions not of their making, nor indeed perhaps of their unmaking, they have shown me by their experience what the theories were, and were not, about. To them I am even more grateful, but I doubt if a recording of their names on these pages is a fitting form of tribute.

This leaves me with a debt to those who, in the preparation of manuscripts, are always landed with the most tedious job, namely the typists – especially Thelma Kassell and Rene Shaw – who have typed the following pages more often than anyone could possibly wish to read them. And then, of course, there was Pauline, who – between commas – did a fine job of the proofs.

University of Sheffield
March 1975

Ankie M. M. Hoogvelt

Introduction

Some years ago, when first encouraged to teach an undergraduate course in the sociology of developing societies to an almost exclusively British student body, my first reaction was 'Whatever for?' And therefore, before accepting the assignment I set out to find some good reasons for teaching such a course. Consulting the *Guide to Opportunities for Development Studies in British Higher Education* proved a not very helpful exercise. This pamphlet, issued by the Overseas Development Institute, claims that 'young people in Britain today are becoming increasingly aware of the plight of the world's developing countries, but are often frustrated in turning their concern into practical action . . . one solution, therefore, is to acquire the skills needed in poor countries'.[1]

It seems to me that such a proposition does *not* present an acceptable rationale for an undergraduate course on developing societies. I do not believe that the twentieth-century type of charitable attitude implied in this statement is realistic, nor do I think that the feelings of idealistic unrest following upon the material saturation so typical of our younger welfare generation should be accommodated for in a 'spell of overseas service' as the pamphlet suggests, however 'compassionate'.

The attitude is not realistic for it ignores the political reality of contemporary developing societies, where attitudes are becoming strongly anti-Western and suspicion of Western expatriates pointing the road to development is growing. Few expatriates that I have met have the inner resources to overcome the stress of such an essentially hostile work situation. More commonly, recourse is taken to social exclusiveness, the cultivation of their own feelings of superiority – indeed bigotry and racism. I remember visiting the campus of a 'bush' school in West Africa, where – in the middle of nowhere and miles from a reasonably sized town – the teaching staff, neatly divided into seven nationals, seven Asian expatriates and seven British V.S.O. workers, had struggled through a whole year of teaching without any inter-racial contact outside working hours.

Apart from having little value as a recipe for international brotherhood, technical-assistance programmes often present a luxury which poor nations can ill afford. The authoritative United Nations Commission on International Development reports that these assistance programmes are anything but the pure 'grant' type of aid which they so ostensibly are meant to be. The Commission points out that, in fact, technical-assistance programmes and their experts have come to represent a considerable financial burden to the recipient country, which is normally required to contribute part of the salaries of the experts, along with such overhead contributions as housing, local transport, office facilities, and so on.[2]

If partaking in overseas technical assistance, and therefore the acquisition of the skills needed in underdeveloped countries, does not present a good reason for our intellectual preoccupation with the developing world, what then does? Let me offer what appear to be two good reasons:

(1) The first reason derives from a 'new look' approach to underdevelopment which has become quite popular in the 1960s although its origin goes back at least a hundred years. Since the last decade, for example, the static mechanical approach to underdevelopment has been replaced by a *structural dynamic* approach. Whereas before underdevelopment was typically seen as a socio-economic *situation* – a situation characteristically lacking in the essential features of the 'developed' situation, namely industrialisation, capital formation, advanced technology, skills, and so forth – today underdevelopment is seen as a *process*, as a structural relation with development. Underdevelopment is now increasingly being understood as the logical companion of the process of development which we have witnessed on our side of the globe. And whereas before the mechanical approach had led to the belief that in order to develop the underdeveloped countries, the rich countries had merely to transfer the 'missing' vital elements, that is capital, technology, skills and certain favourable socio-political institutions, the structural dynamic approach exposes the exploitative, and hence 'underdeveloping' nature of this transfer itself.

Thus it has become a widely accepted thesis that the developed countries – in the past – have been the agents of underdevelopment elsewhere in the world, and that but for the close scrutiny and vigilance of the more enlightened elements amongst us, they will continue in this role in the future.

The recognition of the structural relationship between development and underdevelopment is perhaps one of the most significant discoveries of the social sciences in the last decade. To be sure the origins of this understanding go back to Marx in the nineteenth century, but never before did it succeed in crystallising into some kind of general consciousness, successfully affecting international public opinion and action, as in the last decade.

And now the first effects of this awakening spirit are becoming manifest. For example

(*a*) there is a move, however tentative, towards greater solidarity amongst Third World countries, as for example reflected in the Lima group meetings which were instituted just before the UNCTAD conference of 1972;
(*b*) there is the first powerful formation of a producer cartel, namely of the oil-producing and exporting countries, OPEC;

(*c*) there are serious, sometimes effective, attempts on the part of governments of underdeveloped economies to curb the might of foreign capital through policies of nationalisation and state participation in foreign industries;
(*d*) there is public acknowledgment in the western world that private foreign investment must no longer be counted as a form of aid, and there are consequently pressures at work to increase the 'official' flow of capital resources to developing countries;
(*e*) likewise, there is growing understanding of the counter-productive and exploitative character of so-called 'tied' aid, and a consequent pressure towards multilateralisation of aid programmes.

All these are examples of actions taken or accepted by the international community, which have originated in this new understanding of the structural causes of underdevelopment.

It is precisely because underdevelopment is rooted in the continued capitalist expansion of the advanced nations that both thinking about, and acting in aid of, the disadvantaged nations is an appropriate preoccupation of social-science students in these same advanced countries. To encourage this preoccupation is one of the reasons why this text has been written.

(2) But, one may well wish to ask, why a 'sociology' of developing societies? The answer lies in the phrase André Gundar Frank has coined – 'the underdevelopment of sociology' itself.[3]

There is, for example, a serious difference in both quality and quantity between economic and sociological contributions to development problems. The reason is that whereas economists have long since ceased to regard the national economy as the only economic reality *par excellence*, and have come to consider the economic relations between nations as a legitimate object of their concern, sociologists – with embarrassing naivety – continue to equate *their* object of study, that is 'societies', with what are merely *de jure* sovereign states, and they persist in conceiving of these 'societies' as the only conveyors of social reality. However, as with economic reality, social reality today is *not primarily* manifest or intelligible within the confines of the nation state. Neither 'culture' nor 'social structure' today are particularly societal in character, that is features or 'sub' elements of 'society'. In the 1970s, social structure and culture are first and foremost global in character: for one thing, because the unbridled growth and geographic expansion of capitalist enterprises, coupled with the technological sophistication of communication and transport, have resulted in a scale of *cultural diffusion* unprecedented in world history, and for another because the processes of differentiation and integration of human activities and the subsequent distribution of social rewards such as power and wealth no longer either originate or terminate within ever so many different societies as separate entities. These social structural processes rather run their course at a global level. Social structural extension has accompanied economic expansion and cultural diffusion. There now exists a truly international system of social stratification in which societal values of power, wealth and prestige are distributed from the top of the hierarchy, and spread out in ever so many interconnected though increasingly thin layers of participation across the globe. Thus, the political power of elites in Third World countries is not something that is generated and accorded to them from within their own societies, but rather is dependent upon their position in the global imperialist system and can be regarded as a reward for their contribution – as deputies of imperialism – towards the maintenance of that imperialist system. Similarly, the economic wealth of these elites is commensurate with the degree to which they participate in the global system of production and distribution of goods and services. And finally, even such narrowly cultural values as prestige, that is academic excellence, technical skills and competence, and religious eminence, also derive from cultural definitions originating outside the societies in which they have symbolic and aspirational value.

If, at present, we run into conceptual difficulties in trying to identify these processes and to track them down, it is because sociologists so far have been looking for them in the wrong places. It is, for example, most disturbing to realise that of all the thousands, indeed tens of thousands of sociological volumes written about 'societies' and 'social systems', none, to my knowledge, as yet deals with the sociology of multinational corporations. Yet these corporations directly, or indirectly, organise the lives, the work, the desires and the satisfaction of desires of more people across frontiers than all but the largest and the richest nations do within their own territories. The greater part of world trade takes place not between nations but between affiliates of multinationals. The output of each of the fifty largest industrial companies – at four billion dollars and over – exceeds the gross national product of nearly a hundred member states of the United Nations.[4]

The conceptually most sophisticated, and also the more widely supported school in sociology, namely 'structural functionalism', treats societies as self-contained social systems in which *all* elements and processes of social reality are integrated and obtain their meaning. Ironically it is this very school which is for ever being accused by its opponent, the Marxist school, of adopting a conceptual framework which legitimises and defends the status quo. But how can it, when its conceptual frame does not even describe the status quo? For, to repeat, social reality is not locked up in social systems, called 'societies' and identified with nation-states. But then it follows that there are no social systems, alias societies. Indeed there are none: they are but figments of our imagination, wrapped up in a national flag and accorded a seat in the United Nations.

Following from this it is the contention of this book that to study developing societies as separate and unique entities, each with their own specific cultures and social structures, is a meaningless exercise. Admittedly, the social structures and cultures of societies in three different continents do differ in important and interesting respects, but their *dominant and characteristic* features are similar, deriving as they do from their similar position – both historical and contemporary – in a global relational context.

The Marxist school, to be fair, because it takes its clues for the description of social structure from the economic base, *has* become alerted to the supra-national character of social reality – *vide* its recognition of the structural dynamic relationship between development and underdevelopment. However, the rigour of its other theoretical premise, namely *historical determinism*, does not allow it sufficient scope to accommodate unpredicted change. 'The history of all hitherto existing societies is the history of class struggle' – it is so phrased in *The Communist Manifesto* – but the process of change is at the same time seen to present a *necessary* dialectic: a sequence of thesis, antithesis and synthesis. As long as any one *concrete* historical class struggle has not materialised in its antithesis, that is the revolutionary overthrow of the entrenched social order by the exploited class, history presumably is at a standstill. Thus, for Marxists, it is often difficult to see the working classes in the industrially advanced nations as no longer primarily an exploited class, but as an already exploiting class, that is exploiting both the resources and the artificially created consumer markets of the underdeveloped countries, and in so doing prejudicing the life chances of a Third World proletariat. The social structural extension which has accompanied the economic expansion of the capitalist system of production from the advanced to the underdeveloped areas of the globe, has pre-empted an existing class struggle by turning the original victims of that system of production into its own allies. In this manner historical change has transformed a revolutionary into a reactionary class *without* the 'necessary' overthrow of an entrenched social order.

From a methodological point of view this book presents an attempt to overcome the limitations in the study of development of each of these two major schools by boldly combining their perspectives. The highly abstract – indeed ahistorical – structural-functionalist model of societal evolution will be complemented by a Marxist historical interpretation of international processes of development and underdevelopment. In such a combination of perspectives the problems of contemporary developing societies will appear as dislocations in the evolutionary process which have been generated by the interference and the interpenetration of international capitalism and (neo) colonialism. Seen in this light the structures of contemporary developing societies will obtain sociologically definable 'similarities', so that, hopefully, a case for a general sociology of developing societies can be made.

For practical purposes the organisation of the book will proceed from a working definition of development as 'a process of induced economic growth and change in an internationally stratified world'. This definition contains three focal elements each of which serve as an organising frame for parts of the book:

Part One: *Development as Process*: that is as an evolutionary process of growth and change of man's social and cultural organisation (that is of society);

Part Two: *Development as Interaction:* that is as a process of growth and change of societies under conditions of interaction with other societies; and

Part Three: *Development as Action*: that is as a consciously planned and monitored process of growth and change.[5]

In the first part of the book – 'Development as Process' – we shall be discussing one neo-evolutionary theory in particular, namely the one that has been cultivated by the structural-functionalist school in sociology. There are two justifications for this choice: first, this neo-evolutionary theory recasts a long and authoritative Western intellectual tradition into the frame of an influential contemporary sociological theory; secondly, the evolutionary perspectives of this sociological theory have parented the modernisation theories which, to date, remain the dominant theories about social change and development in the developing countries.

We shall be examining critically these modernisation theories, and our criticisms will prepare the ground for the second part of the book, where development is viewed no longer as a general process of internal societal dynamics, as it is done in the neo-evolutionary perspective, but as a historical process of interaction between concrete societies. Societies are no longer treated as autonomous social systems – which is the methodology of the structural functionalists – but as interconnected parts of an internationally stratified social system. First, the history of international stratification is discussed: we shall trace the dialectics of development and underdevelopment from the period of merchant capitalism through colonialism to the period of neo-colonialism. The theoretical background of this approach will be a Marxist theory of political economy. In this perspective, the so-called Third World countries are seen as occupants of the lower strata of the system of international stratification.

In an effort to combine the structural-functionalist and evolutionary perspectives of the first part of the book with these Marxist theories of political economy, we shall try to interpret the 'typical' problems of contemporary developing societies as ever so many internal dysfunctions due to external penetration consequent upon their enforced entrenchment into the system of international stratification.

The third part of the book, 'Development as Action', finally views development no longer from either of these two points of view, but from the point of view of the pragmatic *planner*. Development is treated as strategy, that is an interrelated set of socio-economic and political decisions, presumably taken or intended by national governments in order to achieve a sustained improvement in the living conditions of the masses of the population. We shall discuss the two dominant alternative strategies, capitalist and socialist, only to end up by discarding both as ideological and pleading for a utopian world development strategy.

Part One

Development as Process

1

Evolutionary and Neo-evolutionary Perspectives

In Part One we shall be concentrating on development as an autonomous process of societal growth and change. When viewed in such narrow terms development comes to be synonymous with 'evolution'. For the time being we shall not bother with the planned management of this process of societal growth and change; historically this is a relatively new phenomenon and we shall only turn to it in Part Three. Nor shall we as yet consider the specific historical context in which the developments of certain concrete societies have been fostered and in which the developments of others have been arrested and retarded. That will be the subject of Part Two. My first concern is with the presentation of an *abstract* and *formal* paradigm for the study of societal evolution.

THE CONCEPTS OF SOCIAL CHANGE AND EVOLUTION

Of all phenomena which are of interest to sociologists 'social change' is perhaps the most elusive and therefore the most given to speculative debates. A survey of definitions and explorations in social change[1] reveals the already considerable diversity of opinion amongst scholars even with respect to such formal questions as 'What constitutes the logical subject of social change?' and 'What are its temporal and spatial dimensions?' Here opinions range from those who identify the subject matter of social change with the entire field of sociology, on the strength of the argument that social life is life and therefore change,[2] to those who prefer to use the word exclusively in connection with alterations in social organisation, and who consequently even exclude cultural change from a definition of social change.[3] Then again, there are theorists who use the concept of 'social change' to denote observable differences in *any* social phenomenon, be it a change in occupational mobility, in size and composition of the population, in school-enrolment figures, or indeed in the political organisation of society, say a change from an absolute monarchy to a parliamentary democracy.[4] Others, on the other hand, reserve the term 'social change' to announce only 'significant' changes in the over-all structure of society.[5] Again, some contributions to the study of social change offer detailed accounts of 'within generation changes in a local community',[6] whereas others present grandiose profiles of the life cycles of entire civilisations.[7]

It is not difficult to imagine that such formal disorientation of the subject of social change has encouraged confusion to prevail in much of the substantive themes of social change as well – substantive themes such as

(*a*) the distinction and the relationship between *micro and macro change.* What are small-scale changes as opposed to large-scale changes? Are large-scale changes the only 'real' changes? Do small-scale changes add up to large-scale changes?[8]

(*b*) the problem of the *continuity or discontinuity* of change. Do large-scale changes of the over-all structure of human societies gradually unfold from an orderly series of small-scale changes, or do they result from crises or abrupt events?[9]

(*c*) the problem of the *causality of change.* What are the causes of social change? Be these changes small or large scale, continuous or discontinuous? Are these causes endogenous or exogenous? In other words, are they germane to the structure of society, or are they external to it? Are they impulses from 'within' or from 'without'? Are these causes material or ideational? In other words, do societies change because of alterations in the material conditions of their existence or do they change because of the emergence of new ideas?

(*d*) the problem of the *directionality* of change. Whether one views social change as continuous or discontinuous, one cannot but wonder if there is some meaningful sequence in the change of human societies through time. Does history make sense? Is there a beginning and an end? Does history run a course? If so, what kind of course; is it a progressive one, a retrogressive one, or, indeed, does history run around in circles?[10]

We shall attend to some of these thorny questions of social change later, more especially in Chapter 3. For the moment we may reflect with a sense of relief that students of development and evolution face a somewhat lighter task than the students of social change, for development as autonomous process, that is as evolution, constitutes but one form of social change the limits of which can be reasonably narrowly defined. For, the semantic meaning of the terms 'development' as well as 'evolution' inarguably introduces the specification of *growth* in the description of change. The word 'development' pairs change with growth; more than this it implies a logical connection between growth and change; and, as we shall see, it explains growth in terms of change, and in turn it explains change in terms of growth.

The word *growth* has a *quantitative* referent only: it refers to an expansion, an increase, a 'more of' whatever it is that one determines to be the subject of growth, be it physical objects, biological organisms or social forms. But the word *change* has a *qualitative* referent: it refers to a difference in the character of whatever it is that one has decided to be the subject of change. By establishing a logical connection between 'growth' and 'change' in the words 'development' and 'evolution' we are inferring that quantitative growth of social life, for example, at some point requires a qualitative change of social life *in order to* sustain and encourage further quantitative growth and change of social life, and so on and so forth.

THEORIES OF SOCIAL EVOLUTION: OLD AND NEW

The viewpoint that the development of social life is a process of continuous growth and change of social life goes back to the early evolutionist school in sociology and in philosophy. Early classical theorists like Spencer, Durkheim, Tonnies, Morgan and many others laboured on precisely this principal feature of evolution, namely that quantitative growth of social life at some stage involves a qualitative change of the forms of social life.

Spencer's theory of evolution, one may recall, may be reduced to two basic propositions.

(1) Both in the development of organic life, as well as in the development of human social life, there has been a process of diversification, that is *many forms* of social life have developed out of a much smaller number of original forms. This, of course, expresses the quantitative aspect of the theory of evolution.

(2) There has been a general trend of development by which *more complex* forms of structure and organisation have arisen from simpler forms. Or, as he elsewhere puts it, there occurs a process of evolution from 'incoherent homogeneity to a definite, coherent heterogeneity'.[11] This proposition acknowledges the occurrence of qualitative change as a necessary concomitant of quantitative growth.

In Durkheim's work, too, we find this notion of an increase, an expansion of social life accompanied by a change in the form of social life. Durkheim argued that when division of labour increases as a result of population expansion (that is the quantitative aspect) then the qualitative character of social life changes also: from one reflecting a 'mechanical' form of social cohesion to one reflecting an 'organic' form of social cohesion.[12] Similarly, Tonnies spoke of an increase of social relationships and a subsequent change of the nature and organisation of social relationships, namely a change from communal to associational organisations.[13]

These older evolutionary models suffered from two main theoretical inadequacies, one being the implicit assumption of uni-directionality and continuity of the evolutionary path. All human societies were assumed to follow a singular particular course between two ideal polar types: from a simple 'primitive' to a complex 'modern'. The most explicit expression of this theorem is to be found in Morgan's work:

> Since mankind were one in origin, their career has been essentially one, running in different but uniform channels upon all continents, and very similarly in all tribes and nations of mankind down to the same status of advancement. It follows that the history and the experience of the American Indian tribes represent more or less nearly the history and experience of our own remote ancestors when in corresponding conditions.[14]

The second theoretical inadequacy of the older evolutionary models is their relative lack of concern with the intermediate stages of evolution on the road from 'primitive' to 'modern'. There were exceptions of course. Spencer, for example, tried to formulate the characteristics of several main types of society – simple, compound, doubly compound and trebly compound – in the evolutionary voyage, but the main concern of the early evolutionists focused on the contrast between the first and the final stage of evolution. And in so doing they have, regrettably, twisted the conception of several generations of anthropologists and sociologists into a stereotypical distinction between primitive and modern man; a distinction which continues to bedevil the development dialogue up to the present day.

The fact that early evolutionary theory wore somewhat thin in the first half of the twentieth century, however, had less to do with these theoretical inadequacies than with the *ideological* bias of the theory. For the core of older evolutionary thinking presented a belief in progress, the belief namely 'that mankind has moved, is moving, and will move in a direction which satisfies ethical requirements', as Morris Ginsberg[15] observes, and he quotes from Spencer's *Principles of Sociology*: 'The ultimate development of the ideal man is logically certain – as certain as any conclusion in which we place the most implicit faith.'

Eighteenth- and nineteenth-century evolutionism expounded an optimistic idea of human progress which the twentieth century, having witnessed two world wars, the atom-bomb, and the ruthless extermination of entire races, could no longer accept.

And yet, so unquestionably useful as an intellectual tool were the main concepts of evolutionary theory, namely those of specialisation and cohesion, of growth and change, that it did not take very long before a *neo*-evolutionary theory emerged, rapidly gaining some popularity. This popularity it owes first and foremost to its blunt avoidance of the 'moral' issue of progress. For neo-evolutionism has replaced 'humanity' or 'mankind' with 'society' or 'culture' as the logical subject of both evolution and progress. The emphasis is no longer on the progress of humanity but on the 'enhancement of the general adaptive capacity of society' as in the theory of Parsons,[16] or on the 'enhancement of all-round capability of Culture', as in the theory of Sahlins and Service,[17] and whilst the assessment of human progress clearly calls for a statement of opinion, the estimation of society's over-all adaptive upgrading can conceivably be done as a matter of objective fact. Indeed the two may even be at odds with each other, as Parsons himself suggests, when, after observing that modern Western societies have a greater generalised adaptive capacity than all other societies thus far, he hastens to add that: 'the adaptive capacity of a society is not necessarily the paramount object of human value. . . . For many people certain aspects of personality, culture, organic well-being, or particular social patterns may be of greater value.'[18]

Sahlins, too, makes the point that progress is not necessarily 'good'.[19] Such observations, of course, already anticipate the core problems of 'Development as Action' with which we shall busy ourselves in Part Three. For it is in the conscious administration and monitoring of the evolutionary process that the question of *choice* between human values on the one hand and greater adaptive capacity of society on the other presents itself.

Which, then, are these objective criteria for assessing 'greater generalised adaptive capacity' or 'greater all-round adaptability'? Here, neo-evolutionists do not seem eager to commit themselves to a wealth of concrete details, but their broad suggestions do run sufficiently parallel to permit us to treat them as in agreement with one another. All stress the relative autonomy of society from the conditioning environmental forces as the main outcome of the evolutionary process. Thus Parsons sees the evolutionary advantage herein that 'forms of social organisation emerge which have increasingly broad adaptive capacities; in their broad characteristics, they tend to become decreasingly subject to major change from narrow, particularized, conditional causes operating through specific, physical circumstances or individual organic or personality differences'.[20] Bellah, from the same School, gives a similar definition: 'Evolution at any system level . . . is a process of increasing differentiation and complexity of organization which endows the organism, social system or whatever the unit in question may be, with greater capacity to adapt to its environments so that it is in some sense more autonomous relative to its environment than were its less complex ancestors.'[21]

From the other neo-evolutionary school, that is from Sahlins and Service, we hear of comparable adaptive advantages. They suggest that evolutionary advanced societies (1) have embodied more varied and more effective means of exploiting the energy resources of a greater variety of environments. Such effective means of exploiting the greater energy resources of more variegated environments means that (2) these societies are relatively more free from environmental control, that is they can adapt to a greater environmental variety than less advanced types.[22] As a consequence, they (3) tend to dominate and replace less advanced types.[23]

Not only is there an apparent consensus on the objective criteria for judging 'progress' or 'generalised adaptive capacity', the neo-evolutionists also seem to be in fair agreement as to the structural characteristics of those societies that display greater generalised adaptive capacity or all-round adaptability. For example, Sahlins and Service suggest the following structural symptoms of general progress: 'the proliferation of material elements, the geometric increase in the division of labour, the multiplication of social groups and subgroups, and the emergence of special means of integration: political, such as chieftainship and the state, and philosophical, such as universal ethical religions and science.'[24] Elsewhere they submit, more abstractly, that societies which have greater all-round adaptability have (1) more parts and subsystems, (2) more specialisation of parts, and (3) more effective means of integrating the whole.

This is not much different from the other neo-evolutionary school, the structural functionalists, who speak of social evolution as a process of *social differentiation* by which is meant the increasing autonomy of the major structures of society (such as religion, polity, government administration, judiciary and economy) on the one hand, and the emergence of new forms of *integration* on the other.[25]

Does this mean that we are back where we started? With the two evolutionary principles of growth and change of the older evolutionary school, namely specialisation and cohesion? Not quite! The replacement of the key concepts of the older evolutionary school by the concepts of 'differentiation' and 'integration' in order to characterise the expansion of social life, its growing complexity and the associated change in the organisation of social life, is not merely a semantic replacement, as Eisenstadt has pointed out, but it reflects an important theoretical advance in the study of society. For it permits an analysis of different stages, levels or plateaux in the evolution of societies as measured first by the extent of social differentiation that has occurred, and secondly, as characterised by the integrative solutions attendant upon each new level of differentiation. In this manner, the neo-evolutionists have made important departures from early evolutionary thinking, and they have overcome at least one of the theoretical inadequacies of their predecessors. Neo-evolutionism now has the tools for the analysis of *various stages* of social evolution. For differentiation, as Eisenstadt says, 'is first of all a classificatory concept. It describes the ways through which the major institutional spheres such as religion, the polity, governmental administration and the economy, have become dissociated from one another, attached to specialized collectivities and roles, and organized in relatively specific and autonomous frameworks within the confines of the same institutional system.'[26] Thus, successive phases of social differentiation each represent a new stage of social evolution. But differentiation alone is not enough. For, if differentiation is to result in 'adaptive upgrading' – and after all that is what the evolutionary exercise is all about – then surely new ways and means must be found to co-ordinate and integrate the newly differentiated substructures with one another, and with the society as a whole, in order that each differentiated substructure may have 'increased adaptive capacity for performing its *primary* function, as compared to the performance of *that* function in the previous, more diffuse structure'.[27]

In other words, each new phase of social differentiation requires a new solidarity and integrity of the society as a whole, with both new common loyalties and common normative definitions of the situation. For example, when in an early phase of social evolution political responsibilities come to be centralised in the hands of a ruling elite, then such structural differentiation will only yield adaptive advantages for the society (say in terms of society's ability to mobilise and organise more effectively a greater number of resources, for instance for public works, military purposes, and so on) when this centralisation of political responsibilities and the ensuing stratification of the social structure – with its vast degree of inequalities – comes to be accepted as legitimate by those who no longer share political power but who are subjected to it. This is what is meant by the problems of integration attendant upon each successive stage of social differentiation. The integrative solution, in this case, as we shall have an opportunity to describe in greater detail in Chapter 2, is that of the cultural pattern of the 'divine kingship'. The godking legitimises his own authority, as well as that of his royal lineage and court retainers, by claiming direct descent from those gods or ancestors who feature prominently in the religious symbol system of his subjects.

Again, at a much later stage in social evolution, the differentiation of the economic sphere calls for entirely different integrative solutions. The separation of the spheres of production and consumption, that is when production no longer takes place in the kinship household, but moves to specialised workshops, factories, or large-scale commercialised farms, and the consequent formation of a labour market, as well as increasing occupational specialisation, call for the institutionalisation of new normative complexes such as a generalised medium of exchange, namely money in association with markets, and the underlying principles of property and contract. We shall return with a more elaborate description of these integrative problems and solutions in Chapter 2. It suffices for the moment to recognise that the neo-evolutionary concepts of differentiation and integration themselves are an 'evolutionary' advance over the older concepts of specialisation and cohesion, because they present a tool for the analysis of successive stages in the evolutionary process.

With respect to the other theoretical inadequacy of the older evolutionary model, namely the naive assumption of the uni-directionality and the continuity of the evolutionary process, the neo-evolutionists have also come up with a rather attractive alternative: evolution by diffusion. Neo-evolutionists make a sharp distinction between the *general* evolutionary process and the evolution of specific societies.[28] The general evolutionary process presents the progression of classes or stages of evolution. The general evolutionary process is seen as an over-all discontinuous accumulation of the evolutionary breakthroughs achieved in the evolution of specific societies. According to Parsons,[29] these evolutionary breakthroughs, or 'universals' as he calls them, *may* be transmitted by the process of cultural diffusion from societies at a higher evolutionary level to societies at a lower one, for whom they then may become a powerful evolutionary force permitting evolutionary advances even beyond the stage reached by the specific society, or group of societies in which the evolutionary universal was first 'hit upon'. In the initial stages of social evolution, for example, *written language* is such an important evolutionary universal, for it encourages the differentiation of the religious and cultural spheres, that is it encourages the emergence of a separate class of *literati* who come to present an authority *independent* of the political authority which was based on the blood ties of a ruling lineage. Therefore written language facilitates the cultural legitimation of authority, and in so doing frees authority from ascriptive ties. *Administrative bureaucracy* which has as its prime feature the institutionalisation of authority in office as opposed to authority in person is another evolutionary universal, presenting – as Weber has argued – by far the most efficient way of organising large-scale operations such as water control, tax administration, mass military operations as well as large-scale productive enterprises. *Money* is yet another example of an evolutionary universal; as a generalised medium of exchange it 'emancipates' resources from all ascriptive bondage, thus permitting free mobility of these resources and so greatly stimulating economic activity. Therefore, just as economic differentiation requires the introduction of money as an integrative normative device (see p. 15), the introduction of money may facilitate and encourage economic differentiation.

It is this recognition of the diffusion of culture patterns of great evolutionary potential which has led neo-evolutionists to conceptualise the *general* evolution of human societies as a discontinuous, a zig-zagging or a leap-frogging process, rather than as a single, neatly definable line.

Talcott Parsons, who, I think, has given the most articulate presentation of neo-evolutionary theory, distinguishes between five successive stages of differentiation and integration, each classing a number of specific societies which he describes by way of exemplifying illustrations:

(1) Primitive societies (the Australian aborigines);
(2) Archaic societies (Mesopotamian empires, Ancient Egypt);
(3) Historic societies (China, India, Islamic empires);
(4) 'Seedbed' societies (Israel and Greece); and
(5) Modern societies (the United States, the Soviet Union, Europe, Japan).

The societies of each stage exhibit a similar degree of social differentiation and have evolved or achieved, or indeed copied from other societies, comparable integrative solutions. It is the *similarity* of the ways of social differentiation and the similarity in integrative solutions of specific societies that is expressed in the term *evolutionary universal*. According to Parsons, an evolutionary universal 'is any organizational development sufficiently important to further evolution, that, rather than emerging only once, it is likely to be "hit upon" by various systems operating under different conditions'.[30]

The adoption of certain evolutionary universals belonging to a 'higher' evolutionary stage, by societies on a lower evolutionary plane, accounts for the leap in the history of specific societies who skip one or more stages altogether. Thus, for example, the fiefdoms of medieval Europe were societies at a lower stage of evolution than their contemporaries, the Indian and Chinese empires. Yet in feudal Europe coalesced some crucial cultural patterns (or evolutionary universals) that had originated in the Roman, Hellenistic and Judaic civilisations, which together combined to transform these medieval societies into modern advanced states. These cultural patterns were, for example, the notion of a 'transcendent god' which wedged a strict separation between Church and State, thus laying the foundations for a secular, rational culture; the concept of 'citizenship' as developed in Roman law, and more radically still in the Greek poleis. Together these two culture patterns provided the basis for the articulation of a formal system of law, which in turn formed the legal matrix for such institutions as contract and property; and these, again, had been necessitated by the growing differentiation of the economy, set in motion (or greatly encouraged) by the introduction of money, and so on and so forth.

The distinction between specific evolution and general evolution, whereby the latter is seen as a product of crucial developmental breakthroughs which have occurred in different specific evolutions, and which have been passed on through cultural diffusion, is helpful in understanding why evolutionary advance in history has taken place *despite* the rise and fall of specific societies and civilisations. It does not, by itself, explain the 'fall' of these civilisations and these specific societies, and neither does it explain their failure to progress further. Sahlins and Service who, in my view, have addressed themselves more adequately to this question, invoke the 'law of cultural stability' to account for the decline of specific civilisations after they have contributed to general evolutionary advance. A 'culture', which in their terminology is no different from what conventional usage prefers to label 'society', Sahlins and Service see as an integrated organisation of technology, social structure and philosophy adjusted to the life problems posed by its natural habitat, and by nearby – and often competing – cultures.

> The process of adjustment or adaptation (i.e. *specific* evolution) inevitably involves specialisation, a one-sided development that tends to preclude the possibility of change in other directions. Therefore, while adaptation (i.e. specific evolution) is creative, it is also self-limiting. A particular technology requires particular social adaptations for wielding it, and conversely a given social order is perpetuated by coordinated deployment of technology. Thus, whereas a given technological development may generate a new organisation of society, the latter in turn operates to preserve the technology that gave rise to it. This we call the tendency toward stability of a culture.[31]

An example that illustrates the operation of the law of cultural stability beautifully, although Sahlins and Service do not mention it, is the historic civilisation of Imperial China. As we shall see in Chapter 2, the system of administrative bureaucracy which had emerged in China as an advanced organisational form adapted to a specific technological need, namely that of the large-scale operation of waterworks, by its very nature presented an insurmountable obstacle to further evolutionary advance, especially with regard to economic development.

Finally, there is one more theoretical advantage to be gained from the neo-evolutionists' distinction between 'general evolution' and 'specific evolution', and that is that it resolves the dispute concerning the question of the material versus ideational character of the causes of social evolution. The recognition of the diffusion of evolutionary patterns enables one to avoid the pitfalls of a mechanistic deterministic explanation of *general* social progress, whilst not denying the causal role of material conditions in the *specific* societies where any one evolutionary breakthrough originates.

DIFFUSION – DOMINANCE AND EXPLOITATION

One important respect in which the Parsonian school differs from the line taken by Sahlins and Service, lies in the respective emphasis on 'diffusion' versus 'dominance' as the process of transmission of social and cultural patterns from one society to the next. In Parsons's approach one gets the impression that the history of mankind has been one happy, relaxed and peaceful exchange of ideas, stimulating progress here, there and everywhere where contact between societies was made. Cultural diffusion appears as a friendly merchant traveller, a timeless Marco Polo, innocently roaming the world, gently picking up a few ideas in one place and harmlessly depositing them in another. Incredulously, the 'domination', 'exploitation', 'imperialism', and 'colonialism', are *not* discussed in any of Parsons's works on evolution.

Sahlins and Service seem to be somewhat more realistic in that at least for them cultural dominance features prominently in their theory of evolution. Diffusion occurs mostly under the wings of dominance. But their law of cultural dominance (note the subtle avoidance of the word 'domination') merely states that each successive higher culture type, because of its greater adaptability, tends, at the expense of lower types, to spread further and faster than previous types 'until in our own day we find that Western Culture is not only extending its dominance over much of this planet but is also attempting to extend into outer space as well . . . therefore the trend is toward a greater convergence and consequent homogeneity of culture types accompanied by a decrease in the diversity of cultures'.[32]

What happens to lower types in the face of this irresistible advance of the higher culture types, in Sahlins and Service's theory, is one of three things: extermination through military conquest; adaptation and absorption of the advanced culture by the lower types so that they become duplicates of the advanced types; or retreat into marginal areas where they may continue their own specific adaptation to the environment for a long time in relative isolation from the rest of the, more advanced, world.

Nowhere in this analysis is there a recognition of the *historical* fact that the most advanced culture type (that is the Western European) has not only spread at the expense of lower types but that this particular advanced culture type has factually *incorporated* lower types as structurally less-developed appendages into its own social, cultural and economic organisation. Literally all that Sahlins and Service have to say about Europe's imperial history (and Parsons does not bother saying anything at all) is that 'the colonial areas had to be politically administered and economically exploited. This meant transplanting many of the political and ideological concomitants of an industrial technology to these societies at the very same time that they were preventing, or retarding, the spread of the technology itself.'[33] There is no analysis of the structural forms of this exploitation. There is no recognition of the historical fact that the social structure and culture of the lower types became reorganised to suit the needs of the advanced types, and in this manner were often *reduced* to still lower levels of evolution. In short, there is no understanding of the phenomenon of *imperialism*. Nor is there a recognition of the historical reality that once the lower types had been joined to the advanced type, as its structural extension, the bond between the two became a life-line for *both*. Nor is there an understanding that the imperialism, or expansionism, of this historical advanced type was, and is, a necessary component of its very structure and culture. And that that is exactly what made it 'advanced' in the first place.

For such analysis we shall have to await the second part of this book. First we must come to grips with what neo-evolutionary theory has to say about the stages of social evolution. Whilst such ordering of themes in this book may be frustrating in view of the fact that the 'imperialist' approach yields a historically more accurate assessment of the processes of development and underdevelopment, it should be remembered that this order of themes is a correct reflection of the chronology of trends in development *studies*, for it is only in recent years that the study of imperialism is slowly replacing the study of evolution. Moreover, an understanding of the differences in stages of social evolution will help us comprehend more accurately the process of structural reorganisation by the advanced types of the 'lower' types during the phase of imperialist expansion of the former.

2

Social Differentiation, Integration and Adaptive Upgrading: The Stages of General Societal Evolution

In Chapter 1 we presented the broad claims and characteristics of neo-evolutionary theories, and we compared these – rather favourably – with those of the older evolutionary schools. The purpose of this chapter is to communicate the substantive theorems of the dominant among the neo-evolutionary theories, namely that theory which has grown up within the so-called 'structural functionalist' school in sociology.

The best treatise of this neo-evolutionary theory is contained in Talcott Parsons's two little volumes, entitled *Societies* and *The System of Modern Societies*.[1] The basis for this neo-evolutionary theory was formed at a seminar on evolution at Harvard University in 1963 to which Eisenstadt and Bellah also contributed.[2] The work of the latter I have tried to incorporate in the following text. My justification for choosing to summarise this neo-evolutionary theory in an introductory text on the sociology of developing societies is that – in my opinion – it presents a concise synthesis of a very long and authoritative tradition in Western European thinking on society and evolution, a tradition to which scholars such as Hobbes, Hegel, John Stuart Mill, John Locke, Auguste Compte, Sir Henri Maine, Herbert Spencer, Emile Durkheim and many, many more contributed,[3] and which tradition was cemented into a forceful intellectual heritage by that most comprehensive of all twentieth-century social scientists – Max Weber.[4] Indeed, if one would want to be unkind to Parsons one might, with good reason, suggest that his work on evolution is no more than an – albeit successful – attempt to recast the work of Max Weber into the mould of a formal analytical framework. But the advantage of this analytical framework, as I see it, is that it permits us to grasp the *essentials* of our own intellectual heritage in a logically compact and concise form.

Parsons's two little volumes on evolution, when written in the late 1960s, were firmly anchored in his general theory of action and his theory of society, as these had been elaborated in great detail in his voluminous *The Social System* which had appeared in 1951, and in the many successors of that work.[5]

Parsons's work had been criticised, among other things, for its unnecessary and incomprehensible jargon (Sorokin contemptuously spoke of Parsonian abracadabra),[6] and his volumes on evolution, which take familiarity with that jargon for granted, have therefore not come as easy reading material. Probably because of this they have not received the credit which I think was their due. At any rate it makes it necessary in the following, first to present a brief and simplified synopsis of Parsons's *theories of action and of society*, before we can move to a presentation of his theory of evolution.

PARSONS'S THEORIES OF ACTION AND SOCIETY: A SUMMARY

Parsons identifies four subsystems of a highly general system of human action: the organism, personality, the social system and the cultural system. These systems are to be seen as each having an interrelated set (hence the word 'system') of mechanisms which control human action. Thus physiological needs, psychological motivations, social norms and cultural values in each of these action systems respectively, guide and control human action. But absolutely crucial to Parsons's action scheme, as developed in his later works, is the idea that these systems of control over human action are arranged in *cybernetic order*, such that the systems located near the top, being rich in information and low in energy, control the systems near the bottom, which are low in information and rich in energy. Thus the cultural system which contains symbolic elements only of knowledge, ideas and beliefs is low in energy but rich in information, and therefore through its information directs and meaningfully orientates the action released by the plentiful but 'mindless' energies of the lower systems, especially the behavioural organism which is at the bottom of the cybernetic hierarchy. The social system mediates and translates the information of the cultural system into prescriptions for action (that is social norms) for the personalities of its participant members, who in turn, by internalising these norms, control their own behavioural organisms.

Social systems are the concern proper of sociologists. Social systems are *systems of interaction* of a plurality of human individuals. The other three subsystems of human action relate to the social system as principal environments which the social system has to articulate with each other and organise within itself. And therefore, although the social system is not the highest in cybernetic ranking, it is yet the most *central* of all four subsystems of action, having the task of integrating all three systems within itself.

Besides these three principal environments, there are, viewed from the position of the social system, two more secondary environments: ultimate reality and the physico-organic world. These are the immediate environments of the cultural system and of the behavioural organism respectively, and they connect up with the social system through their articulation with these two other systems of action respectively. A spatial expression of these relationships would look as portrayed in Figure 2.1 (p. 22).

Society is a *type* of social system, just as universities, the Royal Navy, the Anglican Church or a business corporation are types of social systems. The difference with these or any other systems of human interaction is that society is that social system which has relatively the highest degree of *self-sufficiency*. In other words, society as a

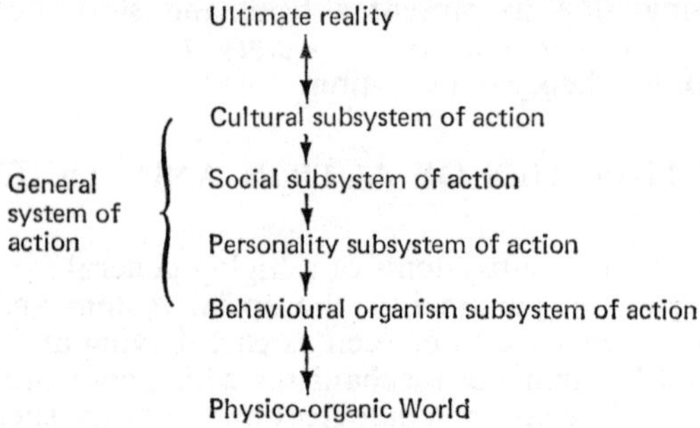

FIGURE 2.1

type of social system 'manages' and integrates its 3+2 environments to a degree unknown to other social systems. By conversion this means that in so far as the units of society (that is subsocial systems like those mentioned above) are segmented, they are more dependent on other units of the same society than on units of other societies.

Social systems, and hence societies may be analysed in terms of four functional requisites:

(*a*) The function of *pattern maintenance*. This function pertains to the relationship between society as social system and the cultural subsystem of action. It entails the maintenance of the highest governing principles of the society, as these are grounded in an orientation towards ultimate reality.

(*b*) The function of *integration*. This consists of ensuring the necessary co-ordination between the units or parts of the social system, particularly with regard to their contribution to the organisation and the functioning of the whole.

(*c*) The function of *goal attainment*. This function pertains to the relationship between society as social system and the personality subsystem of action. It concerns the definition of objectives that are of relevance for the entire society, and the mobilisation of the constituent members of the society for the achievement of these objectives.

(*d*) The function of *adaptation*. This function pertains to the relationship between society as social system and 'the behavioural organism' subsystem of action, and, through it, with the physico-organic world. More concretely it concerns society's more generalised adaptation to the broad conditions of the environment.

THE STAGES OF SOCIAL EVOLUTION

The over-all evolutionary tendency of societal systems is to differentiate into four *primary* subsystems, as each of the societal functions become relatively autonomous and organised in separate subcollectivities and roles (that is culture, judiciary, polity and economy respectively). Greater generalised adaptive capacity of the societal system is the result of this increased autonomy of the operation of these functions.

In the following discussion of these functions I shall describe the developmental tendencies of each of these functions in the process of general societal evolution. Parsons's own division of societal evolution into *three main stages*: *primitive*, *intermediate* and *modern*, as well as his subclassification of societal evolution into *five stages*: *primitive*, *advanced primitive and archaic*, *historic intermediate*, *seedbed societies*, and *modern societies*, will thereby serve to categorise the crucial developmental breakthroughs of the evolutionary process.

Since it is an important claim of this neo-evolutionary theory that each of the four stages of the evolutionary sequel – *after* the primitive stage which is the first and *un*differentiated stage – is associated with the social differentiation of one function in particular, I shall discuss each of these stages in connection with the functional differentiation most critical to it.

Regrettably our scheme cannot remain as simple as that. For *within* each function and *after* its primary differentiation from the societal core, there are further, secondary developments of *internal* differentiation which occur as a result of, and in response to, the primary differentiation of the other functions of society. Therefore, in each section, we shall have to distinguish between the differentiation *of* a particular function and its further internal differentiations. This is particularly complicated, though critical, in respect of the function of *integration*, for this function remains central to the organisation of society at any stage. This is the reason why, superimposed on the classification into five stages, Parsons presents a classification of three *main* stages of evolution, the dividing criteria of which are articulated by the developments of the integrative function. For, in line with the central position of the social system in the general system of action, the societal community which increasingly specialises in the function of integration remains the core of the social system. The critical developments of this integrative function, namely *written language* and *a formal legal system*, are therefore seen as the main watersheds in the temporal sequel of general societal evolution: with written language marking the transition from the primitive to the intermediate stage, and a formal legal system marking the transition from the intermediate to the advanced stage.

Finally, I must again draw the attention of the reader to the fact that we are concerned here with the evolution of society as a *generalised phenomenon*. We are *not* dealing with the evolution of any one specific society. To make the point even more emphatically, not one particular concrete society has ever gone through all the phases of general societal evolution. Societies at the end of the evolutionary sequel are societies that have managed to incorporate in their organisation the crucial developmental breakthroughs (or evolutionary universals) that had been achieved by their predecessors in antiquity and in the great historic civilisations which were transmitted to them through cultural diffusion. To put an even finer point on it, it appears that the one evolutionary universal which more especially has made possible the development of 'modern' type societies, that is the cultural conception of a general moral order – which paved the way for the foundation of a formal legal system, independent of the polity – originated in societies which for peculiar historical reasons never had the opportunity to take full advantage of their own cultural development. I am referring here to what Parsons calls the 'seedbed' societies of Greece and Israel.

THE FUNCTION OF INTEGRATION

The conception of the social system, and therefore of society, as a system of human *interaction* rather than as a system of interacting individuals, has far-reaching theoretical implications. For it means that society, at least in one sense, is distinct from and independent of its constituent members, and as such has needs and problems (of survival) of its own. Such an abstracted notion of society as a reality *sui generis*, as 'a thing in itself', and therefore with properties irreducible to those of its interacting members, is another core feature of Parsons's analytical framework, as indeed it has been for the majority of sociologists since Durkheim. In Parsons's model human social life is seen to take place at two analytically distinct yet again cybernetically ordered levels, that is (1) at the level of individuals, and their actions and interaction; and (2) at the level of the collectivity, of society as a thing in itself.

These levels are cybernetically ordered, such that whatever takes place at the base, that is at the level of acting and interacting individuals, needs to be co-ordinated and integrated at the level of 'society as collectivity'. Seen from this *dual* scheme, the concepts of differentiation and integration can be understood as appropriate descriptions of evolutionary changes taking place at these two respective levels: with differentiation characteristically referring to 'emancipating' processes at the base, that is at the level of acting and interacting individuals who split off into separate collectivities with different roles and interests, and with integration typically referring to the subsequent maintenance problems at the level of 'society as collectivity'. Where integration fails to follow differentiation, segmentation occurs, that is society disintegrates into several smaller subcollectivities, usually reverting back to lower levels of evolution.

In Parsons's analytical scheme, 'society as collectivity' is referred to as 'the societal community', and is regarded as the core structure of society: its primary function is integration. Concretely this involves first, the definition of societal identity and the criteria of membership in the societal community, and second, the normative ordering of the relationships between the members of society, or, at later stages of evolution, between the variously differentiated subunits of society. Since, as we have seen in the first paragraph of this chapter, the social system (and therefore society) is one step below the cultural system in the cybernetic hierarchy, this means by implication that the social system derives the substance of all its norms, including those pertaining to membership criteria, from the cultural system. In short, societal identity is always grounded in a common cultural (or religious) orientation shared by the membership.* In other words, one might say that the societal community defines who are 'we' and who are 'they', and the cultural system says why this is so.

The primitive stage: ascriptive diffuseness

The primitive stage of societal evolution is characterised by the *absence* of any functional differentiation. The societal community embraces one single affinal collectivity only, that is membership in the community is based on one of two principles only: that of descent and that of affinity, relationship established by marriage. Thus the societal community coincides with the kinship nexus.†

In the organisation of the kinship, the normative ordering of the relationships amongst individuals is entirely *ascriptive*, that is differential obligations and privileges are based upon sex, age, descent and affinity, or on any one elaborate set of combinations of these. Besides the obviously biological determination of such a normative order, a grounding principle of the kinship organisation appears to have been the incest taboo, and the resultant preoccupation with the categorisation of marriage prescriptions.

Kinship societies are called 'simple' and 'diffuse', not because they would not be internally complex and elaborate – indeed there may be quite substantial division of labour based upon sex and age – but because there is no functional differentiation *independent* of kinship. There are no functionally specific groups, for example there are no kings and royalties, no class of priests, no administrators, no courts and judges, no occupational/economic groups, and the like. Individuals do not perform different classes of acts in differentiated roles and in different social contexts in the sense true of individuals in more developed societies. All the activities of an individual in the kinship society take place in the same interactive context, that is with the same people around him. A logical corollary of this functional diffuseness is the prescriptiveness of the normative code, that is an individual's ascribed status in the kinship organisation prescribes his behaviour in a wide range of concrete acts, be they productive activities, religious ritual, recreation or whatever. Life in such a society, says the oft-quoted word of Stanner, is a 'one-possibility thing'.[8]

The absence of social differentiation makes for the mechanical type of social solidarity that Durkheim spoke of, namely a type of social

* In Parsons's analysis, the orientational aspect of a culture coincides with what is conventionally labelled as 'religious'. The reason for the conventional terminology is that in the majority of societies at all stages of social evolution, the orientational aspect of culture has involved some conception of 'god', or 'gods'. In those societies where the orientational aspect of culture is focused on metaphysical abstractions which do not necessarily involve 'gods', the function of that orientation is nevertheless the same.

† As Radcliffe-Brown[7] has pointed out, one should really say: *kinship and affinity* nexus, but because there is no inclusive term in the English language for all relationships which result from the existence of family and marriage, the term 'kinship' in anthropology has become shorthand for kinship and affinity.

solidarity which comes from states of conscience which are common to all the members of the same society. Indeed, society 'at the level of individuals' and society 'at the level of the collectivity' are as nearly one and the same as they will never be again. In primitive societies, integration is the most salient of all societal functions; the execution of the other functions is controlled by the exigencies of ascriptive solidarity. Thus, religious ritual, the exercise of political authority, the enforcement of law, and even the provision of material needs are organised so as to promote the social solidarity of the community. The kinship organisation, as an integrative device, is a first evolutionary universal in the temporal sequence of human societal progress. Since, as stipulated before, societal identity is always grounded in a common cultural/religious orientation, we see at this stage of societal evolution the kinship organisation reflected in the religious symbol system of primitive man. The religious orientation of primitive man is towards a mythical world of ancestors, some human, some animal, who although in some respects are different from, and more heroic than, ordinary men, are never completely distinct from ordinary men. The mythical world is, rather, an exemplary world, in which all the routines of daily life are prefigured and all the details of concrete human relationships within the kinship organisation are pre-cast. By acting out the myths, in religious ritual, every member of the community can participate in this world of ancestors and archetypes, and fully merge with it. In such a belief system there is no need for separate religious organisations, priests or cults. The lack of differentiation between the religious and empirical worlds reflects the lack of differentiation within the kinship organisation.[9]

Further developments in integration – inclusion; written language – a formal legal system

At later stages of social evolution, and consequent upon social differentiation, the societal community will develop an increasingly stratified and (later still) plural social structure, with different subcollectivities occupying different positions, and having different rights and duties. Time and again such stratification will require (1) a new definition of societal identity, and (2) an increasingly complex normative system in order to regulate and co-ordinate the relationships amongst the differentiated subunits.

The general evolutionary tendency thereby is toward the definition of increasingly higher-order generalised collectivities, and the subsequent *inclusion* of the differentiated stratified units into the 'relevant' societal community for each stage of social differentiation. More concretely, in social evolution, the particularistic basis of membership in the societal community is progressively being replaced by more universalistic criteria of participation and membership. Taking the three main stages of social evolution as the three main points of reference, there is a progression from 'kinsman' through 'subject' to 'citizen'.

With respect to the normative order of the societal community, two evolutionary universals have developed which have served to mediate and sustain the increasing complexity of the social order through the stages of societal evolution: *written language* and a *formal legal system.*

It may seem strange to consider written language an 'integrative' device which permits substantial specialisation of activities in a society. Yet that is exactly its main virtue and this is why scholars through the ages have looked upon written language as the dividing criterion between the primitive and 'civilised' worlds. For written language permits the structural differentiation between the cultural and social spheres of society. It makes the cultural system relatively autonomous *vis-à-vis* the societal community, and so permits the accumulation, entrenchment and continuity of a cultural tradition. It acts as a stabilising force since such a tradition no longer depends on the fallible memories or idiosyncracies of individual bearers of culture, and, written codes, by allowing communication to take place in non-interactive contexts, co-ordinate and stabilise social relations between specialised subunits of the society. In the initial stages of societal evolution when literacy is the prerogative of a small upper class of religious and/or political elites, it helps to license the authority of these elites over the mass of the illiterate population. But in modern societies, where literacy has been extended to the entire population, it greatly furthers the participation of the entire population in the polity. This is the reason why literacy is usually hailed as an important tool in achieving 'democracy'.

If written language presents the 'watershed' between the primitive and civilised worlds (which commences with the intermediate stage in societal evolution), a formal legal system with its corollary, an independent judiciary, is the hallmark of modern societies. Its merit as an evolutionary universal lies in the fact that it permits the separation of a society's integrative function (that is co-ordination and regulation of specialised activities) from political authority. More particularly it 'frees' the economic/occupational sphere from the clutches of the polity, and from the political and/or religious supervision to which it was subjected during the intermediate stage of social evolution. The formal legal system, which modern societies owe to the genius of Roman law, is *calculable* to a high degree: it does not depend on petty interpretations based on ritualistic/religious or magical considerations, nor on that of politically powerful personalities. Fundamental to the formal legal system is a set of universalistic values which are seen to apply to *all* members of the society independent of their status in the social hierarchy. Only when a society has such an independent, highly generalised and formal legal system at its disposal can it effectively develop and administer the legal codes of 'contract' and 'property' upon which the full emancipation of the economic sphere rests.

THE FUNCTION OF GOAL ATTAINMENT

The function of goal attainment pertains to the relationship between the societal community and the personalities of its constituent members.

First of all a society needs to develop and maintain adequate levels of motivation on the part of its members for participating in socially valued and controlled patterns of action. This aspect of the goal-attainment function becomes crystallised in the *socialisation* process. At all times, and in all societies, however, the socialisation process remains factually incomplete, so that additional mechanisms of social control, more especially of *enforcement* are needed.

Secondly, society needs to organise collective action for the attainment of collectively significant goals, be it with regard to the defence of its territory *vis-à-vis* other societies or be it more generally in terms of the organisation of collective action to meet certain conditions in the physical environment. Thus the relationship between society and its constituent members (or units) becomes the reference point for what is conventionally labelled as the *political aspect* of society.

The structural differentiation of the polity – vertical differentiation – the advanced primitive stage and archaic societies

In the history of general societal evolution, the political function, including that of enforcement, is generally considered to be the first function to 'split off' from the 'seamless web' of primitive kinship organisation. Many different causes for such evolutionary development can be put forward, but a likely one is that certain alterations at the basic setting of human existence, such as population pressure, or important technological innovations (for example settled agriculture) put a premium on permanence of location and therefore on place of residence. Consequently the need arises for internal distribution of land of different economic advantage. It is at this point that a process of vertical differentiation, or stratification, sets in: existing ascriptive prestige differences between lineages in the wider kinship unit come to articulate differences in economic advantage as well.

This process of vertical differentiation tends to go hand in hand with the structural differentiation of the political function away from the societal core: naturally, the economically advantaged groups tend to usurp political responsibility in an effort to consolidate their position, and thus the seamless web turns into a two-class system of stratification. Such stratification Parsons appreciates as an evolutionary universal since it presents a first and necessary step away from the ascriptive and prescriptive rigidity of the kinship organisation. The members of the upper class are sufficiently advantaged and secured to be willing to take risks and to initiate important changes. Moreover, they are now in a position where they can requisition resources, physical as well as human, that otherwise would not have been released, for new forms of collective action. However, this social differentiation creates problems of integration: the new social order with its two-class system of stratification needs a new definition of societal identity, a new criterion of membership. Otherwise the two classes might as well simply split up altogether and segment into two separate kinship societies. Indeed, the reader should remember that such segmentation at this stage of societal evolution is the more frequent phenomenon. Consolidation of the two-class society is only then accomplished when the *principle of domicile* becomes included gradually superseding the criteria of membership by blood and marriage. Thus the new societal community is one whose membership criteria are defined by both kinship and territorial boundedness. At the same time this new definition of societal identity provides the basis for the legitimation of the advantaged position of the upper class. Redefinition of societal identity is always rooted in a reorganisation of the cultural system, that is, at this stage – the religious symbol system. The upper class, which monopolises political control over the territory and the entire societal community, legitimises its advantaged position by claiming direct descent from those progenitors, or ancestors, who, as we saw earlier, inhabit the 'mythical world' of the religious symbolism of primitive man. Ancestors then come to be conceived as the founding fathers of the community: those who arrived in that territory first and who decided to settle there. Legitimation by consanguinity with the ancestors permits the upper class to claim and cultivate a special talent for mediating and communicating between the ancestors and the ordinary men (that is the lower class). In other words, at this stage of social evolution the most probable occurrence is for those who monopolise political responsibilities and superiority to claim religious responsibilities and superiority as well. At this stage of social evolution, therefore, we see the splitting off of religico-political activities from the main societal core: they become socially and culturally distinct activities (although not distinct from each other yet) carried out by a specially designated subcollectivity, for example *an upper priestly class.*

The special position of the upper priestly class in relation to the ancestors, and their claims at special talents for communicating with them, leads by implication to a conception of a far greater *distance* between ancestral figures and 'ordinary men' than had been the case in the religious conception of primitive society. The ancestors of primitive religion turn into *gods*. Gods, who are very distinct from ordinary men, far more powerful and far more superior, *control* the world of ordinary men, wilfully but arbitrarily. Consequently man needs to worship the gods, make sacrifices to them, and generally pacify them into benevolence. The thus created gulf between gods and ordinary man puts a special premium on those who can rightfully (in terms understandable within the context of existing patterns of social organisation, namely ascriptive terms) claim to be in a position to bridge that gulf.

Thus the integration of society, which has become differentiated into a two-class hierarchy in its ultimate form, becomes stabilised in the cultural pattern of the 'godking', who with his lesser but still divine lineages has ruled over so many societies in all parts of the world. We encounter the pattern of the divine kinship both in the political kingdoms of *the advanced primitive stage* (especially African kingdoms such as those of the Nupe, the Shillouk and the Zulus) and in the *archaic societies*, the principal subtype of the intermediate stage of societal evolution (for example Ancient Egypt, Mesopotamia, the Aztecs, Incas and Mayans). What distinguishes the two is a degree of differentiation of the religious from the political sphere, which in the archaic societies is furthered by the development of written language and the ensuing craft literacy amongst the priesthood. In fact, archaic societies exhibit a *three-class* system: the divine king and his royal lineages, a middle class of priestly officials who in various degrees participate in the charisma of the king, and who in his name administer not only temple rites but a wide variety of bureaucratic tasks, and finally a mass of illiterate peasants and artisans.

Further developments in the polity; bureaucratic state administration – democratic association

Still further differentiation *within* the political sphere itself, notably the separation of effective power from authority, accompanies the transition from the archaic to the advanced intermediate stage. *Bureaucratic administration* which institutionalises *authority in office*, and which makes a clear distinction between 'office' and other aspects of the personal status of the incumbent, is a second evolutionary universal in the political development of societies. The bureaucratic organisation greatly enhances the organisational effectiveness of the polity because it permits the exercise of power by a great many people who neither need personal legitimisation for their authority, nor personally (ideally of course) benefit from it. Thus, in the bureaucratic administration, power becomes a 'neutral' symbolic medium of mobilisation and acquisition of resources in the interest of the collectivity. Indeed, the bureaucratic administration is indispensable in societies where either the size of the population, or certain conditions in the environment, require mass mobilisation for large-scale operations. A prerequisite for bureaucratic administration, however, is the completed differentiation of the cultural from the political spheres which we shall discuss below.

Finally, in modern societies, argues Parsons, the completed social differentiation of *all* functional spheres, including the economic/occupational sphere, has created such a multi-stratified, plural system of society, that effective exercise of political responsibilities can no longer be accomplished without the *mediation of consensus*, both in policy formulation and in the exercise of power, on the part of all the differentiated subunits of society. *Democratic association* with its features of fully enfranchised membership and elected leadership is therefore a final evolutionary universal and belongs to the modern stage. It differentiates – within the polity – power from leadership, giving the power to those who are led.

THE FUNCTION OF PATTERN MAINTENANCE

This function of society pertains to the relationship between society as social system and the cultural system, which is the action system highest in cybernetic ranking.

Cultural systems exist by virtue of the problem of 'meaning' which is a problem exclusive to the human race. Cultural systems are basically sets of interrelated answers to fundamental questions about the human condition. Questions like: Who am I? Where was I before I was born?

Where am I going after death? What is real, what is unreal? What is true, what is false? In short they are questions about the ultimate conditions of human reality, about 'ultimate reality'.

Man, in order to act coherently, must have some idea as to the sense of his being and as to the sense of being in general; for asking about the sense of one's being implies questions about the justification of one's actions; about the right way of doing things. The fact that man throughout history has needed to raise these same sort of questions about the whys and wherefores of human existence, explains Parsons's viewpoint which treats ultimate reality as an environment external to the action systems (including the social system and thus society), and in this respect, therefore, as comparable to the physico-organic world. Both are action-exogenous environments. But whereas the physico-organic world is the basic *conditioning* referent of human social life, man's orientation toward ultimate reality is the highest *organisational* referent of human social life. And, as we shall see, man's freedom from the narrow conditioning factors of the physico-organic world depends to a very great extent on his reorientation toward ultimate reality.

Although societies throughout history have raised essentially the same sort of questions about the human condition, the answers have been very different, which explains the great variety of cultural systems in the history of mankind. But each time, these answers, or commitments to ultimate reality (or 'religious orientations' as they are conventionally called) have provided societies with the building blocks for their social order, and have served as guidelines for man's action in the physical world.

Parsons's terminology is confusing in that we encounter cultural systems both outside and inside the society. Yet a moment's reflection will reveal the reason for this double terminology. For there are cultural systems which form an integral part of many different societies, providing the legitimation for their patterned normative orders, and meaningfully orientating the behaviour of their members. Thus, for example, Islam, Hinduism, Buddhism, Christianity, Science and Scientific Socialism, are such cultural systems which cut across the boundaries of many societies. At the same time, and inversely, we can also identify (and, through archaeological and ethnographic work, reconstruct) cultural systems which are no longer integrated into existing societies, as for example the 'dead' civilisations of Ancient Egypt and Mesopotamia, just as we can also identify cultural systems which exist merely as 'revolutionary' ideas in the minds of certain personalities or subcollectivities in the society and which have not yet crystallised into foundations for a new social order. It is only when a cultural system becomes institutionalised within the social system that we can speak of the cultural (sub) system of society.

Thus society as social system translates beliefs and cognitions about ultimate reality into cultural *values* underpinning its social order, and orientating man's behaviour in the physical world. And this is the reason why Parsons places cultural systems at the summit of the cybernetic hierarchy of systems controlling human action. But this, of course, also has tremendous implications for his theory of evolution. For it means that we must focus on the cultural system in order to detect the major sources of general societal evolution, and since evolutionary advance is associated with the relative degree of autonomy of each of the primary functions of society, we can expect the *major* evolutionary breakthroughs to be associated with the progressive differentiation of the cultural system away from the societal core.

The directional tendency of cultural evolution is towards a *generalisation of the values* operative in cultural systems. Such value generalisation is in keeping with the need to legitimise the progressively wider circumference of the societal community, and the need to provide value patterns that may regulate the activities and the interactions of increasing numbers of differentiated subcollectivities, each with their own specific goals and means for achieving them. Thus, in general cultural evolution we have come a long way from the prescriptive codes operating in primitive societies, where myths chart every minute detail of human existence, to the abstract, highly generalised value patterns which direct our own behaviour and integrate the plural social order of our own modern societies, for example values such as 'equality', 'freedom', 'brotherhood', or the maxim that every individual has the right to life, liberty and the pursuit of happiness, which is written in the American Constitution, or the maxim, *pacta sunt servandi* (contracts should be honoured) which is written in the constitution of just about every modern society. Indeed, one American philosopher once observed that 'the Constitution is made for people with fundamentally differing opinions'. It is because of these differences amongst the individuals in modern complex societies that the moral values underpinning their social order need to be highly general and *unspecific.*

Because the values of the cultural system derive from beliefs and cognitions about ultimate reality, a generalisation of such values must somehow reflect a fundamental 'rethinking' about ultimate reality. Broadly speaking, as Bellah well argues,[10] this 'rethinking' in religious/cultural evolution has involved the conception of *growing distance and separation* between ultimate reality and ordinary, empirical human reality.

In primitive religion, we observed, the orientation towards ultimate reality is contained in a conception of a 'mythical world', inhabited by ancestors who are not very different from ordinary men, and whose acts are not very different from those of ordinary men. Indeed, the mythical world is a paradigmatic world, a blueprint for this world. In the eyes of primitive man, it is the only *real* world. Human acts and physical objects in the empirical world are only real to the extent to which they participate – through religious ritual and ceremonies – in the mythical world. 'Real' are only those acts which are an imitation or a repetition of ancestral archetypes, and 'real' are only those objects in the external world which have a mythical 'meaning' – to primitive man, being is meaning. Out of a countless number of stones which to him are meaningless and therefore beingless, primitive man will select one, and endorse this one with meaning – and hence being – for this one stone is, for example, the place where the ancestors lay buried.[11] Thus, in primitive religion, the boundaries between the ordinary human world and the mythical world (that is the world of ultimate reality) are blurred and they disappear altogether in religious ritual, which at this stage of evolution is not a prerogative of any one particular class of priests, but a communal activity in which everyone participates.

Following the first process of social differentiation, namely the rise of a separate politico-religious class, the mythical world transforms into a *supernatural world*, housing 'gods' not ancestors; and these gods are no longer merely exemplary to men, they are *in control over* men.* Moreover, there are many gods, indeed there may be as many as there are rival priestly groups and cults, and their control is arbitrary, whimsical, unpredictable, and hence needing special means of negotiation, for example sacrifices and worshipping. The functionally specific class of priest claims special talents in bridging the gulf between gods and men, and thus human destiny and human activity becomes a matter of revelation and divination, of traditions, texts and scriptures, handed down by the gods to those whom they favour (most frequently their relations on earth – as in the pattern of the divine kingship). In these archaic religions, there is still a clear connection between the human and the supernatural world. The human world is seen as part of a natural divine cosmos. The traditional social order is anchored in a divinely instituted cosmic order, and social patterns are enforced by religious sanctions at every point.

The structural differentiation of the cultural system – the historic† civilisations of the advanced intermediate stage

The link between the world of 'ultimate reality' and that of 'human empirical reality', is completely severed with the appearance of the great historic religions of Islam, Buddhism, Hinduism and Judaism. Common to these religions is the conception of a supernatural world that altogether *transcends* the human world. It is the domain of the 'sole creator of the universe', the one 'god', who has neither court nor relatives, and whose virtues and attributes are infinitely beyond what is distinctive in the human realm.

The evolutionary importance of this conception of a 'transcendent

* The following is mainly adapted from R. Bellah's contribution to the aforementioned symposium on social evolution. Bellah's essay deviates from Parsons's work in that he identifies different watersheds between the main stages of evolution. This is because Bellah concentrates on religious evolution, whereas Parsons's focus is on the integrative sphere of society. Thus, whereas for Parsons the main dividing criteria between the stages of social evolution are 'written language' and 'a formal legal system', for Bellah the main watersheds lie in the religious sphere: mythical world – gods – 'transcendent god' and the Reformation.

† The term 'historic' in this context refers to the participation of all the adult population of the upper classes in the literary tradition. This is distinct from archaic societies where literacy is 'craft' literacy and restricted to a special class of priests.

god', and of a 'transcendent supernatural realm' altogether, is that *it prepares the ground for universalistic value patterns.* For, from the point of view of these religions, it no longer matters who a person is, what ethnic group, clan, or family he belongs to, or indeed which elevated position in the social order he may have. For, in the eyes of the transcendent god, all men are equal and all men are, in principle, capable of salvation. The dualism between this world and the transcendental 'other' world in historic religions, however, has two further implications: first, the discovery of another world infinitely superior to this world is accompanied by a devaluation of this world and this-worldly affairs. Religious concern from now on concentrates upon the 'spiritual' realm and denounces the 'material' realm. World rejection becomes a necessary condition for religious salvation. At the same time, and secondly, the universalistic notion that all men are equal in the eyes of 'god', opens up the *theoretical* possibility of religious salvation and merit for all. In the social sphere this has the double consequence of, on the one hand, encouraging the emergence of a religious elite *independent* of the political elite, which is a step forward in evolutionary terms, and on the other hand of frustrating development in the material sphere, notably economic and technological. For the social structure in any society reflects its dominant values, and 'world rejection' as a dominant theme did not exactly encourage the rise of merchants, traders, craftsmen or artisans in the social hierarchy. They were second-class citizens and few who could afford to avoid it would seek to join their ranks.

A comparable dichotomy between ultimate reality and human reality, and with roughly the same implications, is characteristic of the *speculative philosophical traditions* of Confucian China, of Ancient Greece, and of the Roman Empire. Of course, there were vast differences between all these civilisations; differences that partly stemmed from the degree to which either one or both of the two implications (that is of world rejection and of 'equality') were institutionalised into the social order – with India probably going furthest in institutionalising world rejection in the code of Dharma, and Rome going furthest in institutionalising 'equality' in its legal concept of 'citizen'. And yet, with the exception of Greece and Israel, Parsons classifies all the remaining four (Imperial China, Imperial India, the Islamic Empires and Imperial Rome) as belonging to the same *advanced intermediate* stage of societal evolution, for in all these four societies the sharp differentiation between ultimate reality and human reality led to a *new dualism* in the social order.

In contrast with, and advancing beyond, the type of social order in archaic societies, the upper class of the historic societies could no longer claim authority or high status on the basis of some divine connection or ascriptive descent. *Rather, in these civilisations, the status and the authority of the upper class was defined in cultural/religious terms.*

Knowledge of, and action in accordance with, the conception of the ultimate truth, or ultimate reality, became the qualifying attribute of a new class of 'superior men'. This cultural legitimation was most marked in China where it became fully institutionalised in the examination system.* Such cultural determination of the social order presented an evolutionary advance over the archaic type of social order because it considerably broadened the basis for entry into the 'relevant' societal community by introducing some universalistic criterion of merit or achievement. It permitted the separation of legitimation of authority from the execution of authority. Furthermore, the meritorious and competitive position of the scholar/official fitted in neatly with the requirements of China's rational bureaucratic system of administration.† For the principles of rational administration based on specialist knowledge and the employment of experts demanded the broadest possible social base for recruitment. And yet this universalistic basis of the social order in Imperial China was not broad enough to successfully incorporate newly differentiating elements, namely the rising urban trading and manufacturing classes. Since, in the Chinese system, these classes could not obtain 'full-membership status' they were thwarted in their aspirations and in their potential contributions to further societal growth. Deprived of full legal rights, they were an easy prey for the exploitative and extortionate practices of the 'superior' men.[12] At the same time, because the system was universalistic enough to allow, *in principle*, anybody to attain the privileged position of scholar/official, provided he was qualified through education (which as it happened was a costly affair), it deflected ambitious and competent individuals away from commerce and industry. Thus, because the status of the 'relevant' societal community was defined in cultural terms, the organisation of the system did not allow sufficient scope for differentiation and development in the economic sphere. This cultural *stabilisation* not only 'precluded' adaptation into different directions but it ultimately, by its own internal processes, led to stagnation and decay, since whatever economic advance there was went into expensive education for ever more aspiring bureaucrats. In the later years of the empire,‡ there was an oversupply of prospective bureaucrats who were with qualifications but without function. They were a transitional group between commoners and those qualified to hold office. Frustrated, they became a source of revolt and insurrection

---

* The fact that the Chinese system included an emperor and a court of retainers who were in a sense 'above' the bureaucracy, and were exempt from qualification by examination, did not – according to Parsons – effectively influence the *cultural* legitimation of the social order of Imperial China, as described here. For the Chinese emperor held only a rather vague 'mandate of heaven'. He was a kind of Pope, and although he held a divine mandate, he was himself not divine.

† Imperial China's bureaucratic system of administration may well be described as a 'rational bureaucracy' in the Weberian sense, for it fulfilled the three criteria of rational bureaucracy, namely strict adherence to formal rules; rigid separation of personal from official status; and an internal hierarchy of authority with precise determination of spheres of competence.

‡ We are referring here specifically to the last empire, although one may well suggest that these internal processes of stagnation and decay had probably contributed to the fall of successive earlier empires too.

within the system. At the same time, and by the same internal process, the system would produce too many *employed* bureaucrats as well, and at ever lower salaries. This would encourage more abuse and corruption on the part of the lowly paid bureaucrats, whilst the fact that ever more bureaucratic parasites would need to live on the surplus generated by the peasantry would lead to ever more frequent revolts.

The structural differentiation of the judiciary: The seedbed societies of Israel and Greece

In terms of cultural development, *the seedbed societies* of Greece and Israel went furthest in value generalisation, thus permitting the conception of a corporate body of citizens and the potential inclusion of all classes and groups into the societal community. Paradoxically this was because both cultures developed a conception of a moral order governing human affairs *totally independent of any particular societal or political organisation.* The development of a formal and independent legal system (as institutionalised to a degree in both Greece and Rome), in which members of the corporate citizen body had equality of the franchise and other *basic rights before the law*, derived directly from this notion of a transcendent and international moral order. The limitations of the Hellenistic and the Roman civilisations were, however, again in line with those of the other historic civilisations in that the relatively egalitarian corporate community, the citizen body, was only seen to pertain to an upper privileged class within a larger system, much of which was denied full rights of membership (for example slaves, resident aliens, different ethnic groups, 'barbarians', and so forth). The Judaic civilisation potentially went further: it added to the notion of an independent moral order the idea of inclusion into the community of everybody who voluntarily obeyed the 'will of god' (in other words, voluntarily adapted to that independent moral order). Thus, for the first time in the history of societal evolution, the onus of membership lay with the individual's *will.* Membership of the relevant societal community became a matter of *voluntary* association.

Because of peculiar historical circumstances of conquest and dispersion, Israel was prevented from practising what it preached, at least in its own territory. But the message sunk in and travelled with Christianity to Western Europe, where it helped formulate an absolutely unique reorientation towards ultimate reality, in the Protestant Reformation.

Further developments in the cultural system – secularisation

It is only in the Reformation that the notion of equality before the eyes of God becomes fully institutionalised in the social order. For the Reformation kept the notion of a transcendent god and that of duality between this world and the ultimate 'other' world, but it abandoned the idea of world rejection as a condition for redemption. On the contrary intensive activity in the world, provided it was done to serve the 'glory of God', became the road to salvation. Because the Reformation put the emphasis on 'faith' as an internal quality of the individual, rather than on particular acts clearly marked as 'religious', salvation from now on was *literally* and no longer just theoretically within the reach of 'everyman'. Carpenters, clockmakers, butchers, traders, merchants and peasants as well as priests, scholars and monks could *demonstrate in their daily duties* that they belonged to the 'chosen people'. In short, the cultural orientation in Western Europe from the Reformation onwards legitimised the inclusion of men from *all* walks of life as first-class citizens, and this, of course, by implication reduced the preference for religious functions as the only respectable social positions.

But the Reformation did more. By emphasising intensive activity in the world as a means to serve the glory of God, it also radically stimulated the differentiation and development of the 'adaptive function' of society, to which we shall turn next.

THE FUNCTION OF ADAPTATION

This function pertains to the relationship between society as social system and the behavioural organism, and through it to the physico-organic world.

For 'man' in society, the most immediate and tangible environment is the physico-organic world. Both the physical-geographic location in which man finds himself and his own physiological characteristics and biological needs are environments with which he has to 'cope', to which he has to adapt, in order to survive. In this way, the physico-organic world consitutes the basic conditional referent for human life.

It is important to remember that the external environment constitutes but a conditioning, not a determining factor in human life. Thus, man's biological needs and physiological characteristics are mere organic foundations which limit the possible range of behavioural responses; they do not determine the actual content of his behaviour. For the actual content of the behaviour pattern is generated by man's culture and man's social organisation, that is society.

It is true that man has certain innate needs such as the need for food, for shelter, for sex, for affection, for avoiding pain, and so on. These needs are biological characteristics common to all mankind. But these characteristics do not determine the behavioural response to which they give impulse. They are mere bodily states which initiate tendencies to a general *class* of activities; they do not specify the activity. There is, for example, tremendous variation between cultures in the definition of 'food'. People in cultures different from our own eat many foodstuffs which we would not recognise as food, which would never arouse our hunger-drive, which would never 'make our mouths water' as the saying goes. Moreover, food habits, in turn, affect our organic foundations, notably our digestive system, so that it is doubtful if a European could survive on a diet customary to the Australian aborigines, and vice versa. Further still, society not only shapes our organic needs, it also penetrates into and *transforms* our physical characteristics. Increased calorie intake in European, and recently also in Japanese societies, has increased the average height of their inhabitants. Improved nutrition and medical aid has helped to ward off diseases and generally strengthen the organic base, with the result that the average life span has been doubled.

We can now understand why the behavioural organism constitutes an 'environment' to society as social system. Serious problems of integration and management occur at the societal level, when the behavioural organisms of the members of society have become orientated to ways of need satisfaction which transcend a society's adaptive resources. This happens to be exactly one of the most conspicuous problems in developing countries today. It connects with the more familiar problem of the 'revolution of rising expectations', when, after independence and because of several centuries of European contact, the local populations in these countries demanded the material forms of human existence that were only available to the people in more developed countries. The result has been a complete breakdown in the 'self-sufficiency' of these so-called developing societies. We shall have occasion to look at this problem again in Part Three.

In ways comparable to those of man's organic foundation, we can regard man's physical-geographic location as a conditioning rather than as a determining force in human life. The natural conditions of the locality of a society to an extent do indeed necessitate particular kinds of human activity, and these in turn give rise to particular forms of social and cultural organisation. Anthropologists observed long ago that the surrounding in which a community is placed determines its primary interests, and that those in turn affect the entire character of its vocabulary and the make-up of its social system. It is not surprising that the Eskimos have seven different words for snow, or that desert cultures like those of the Berbers revolve around camels with no less than forty expressions for the beast. Yet, here too, more so than is the case with man's organic base, we observe that man's socio-cultural intervention reshapes and alters the environment, and in so doing further diminishes the importance of the environment as a conditioning factor in human social life. The invention of agriculture makes man less dependent upon the natural richness of the soil, and hence permits settling in relatively inhospitable surroundings, and with industrialisation we move further still from nature's restricting grasp.

The adaptation to and the interference with the external environment is the function of that culturally and socially organised human capacity which we call *technology*. In Parsons's analytical scheme, technology comprises all those procedures, skills, crafts, techniques and tools ranging from primitive hunting and fishing gear to modern electronic computers, which together permit the active controlling and altering of objects of the physical and biological environment in the interest of some human want or need. And when neo-evolutionary theory defines the result of general societal evolution in terms of an increase in society's generalised adaptive capacity, it refers in the first instance to the increasingly dominant position of technology *vis-à-vis* the external environment, which reduces man's dependence upon it. More advanced societies are more autonomous relative to their external environment than less-advanced types. The more technology encroaches upon the environment, the less technology and society are conditioned by it. Whilst it may be a fruitful exercise to explain the culture and the society of the Eskimos in relation to the Arctic conditions in which they find themselves, it is no longer instructive to explain Dutch society with reference to its location in the low lands by the sea.

Not only in Parsons's, but in practically all theoretical models of society, technology is seen to occupy the lowest rung of the socio-cultural ladder with which society raises man from the bondage of his physico-organic world. But the word technology really only refers to the instrumental and behavioural forms of the adaptive function. The organisational contexts of that adaptation, more concretely the allocation of resources, the assignment of work-roles, the co-ordination of productive activities and the distribution for consumption, are conventionally labelled *the economy*.

As was the case with the other three main functions of society (that is of integration, of goal attainment and of pattern maintenance) in the general evolution of societies, the adaptive function, too, differentiates into separate subsystems, namely those of the economy and technology.*

The history of general societal evolution suggests that the full differentiation of the economy and technology only occurred after each of the other three functions had obtained a degree of relative autonomy and independence both from each other and from the societal core. In other words, economic and technological development as we have witnessed since the commencement of the 'modern' era in Western Europe was preceded by the political, cultural and judicial developments that we described in the earlier paragraphs of this chapter. Neo-evolutionary theory, however, claims that these developments were not merely *historical* antecedents, but that they are *logically* necessary requisites for economic and technological development. Since these claims have such very far-reaching implications for contemporary theories of modernisation, we shall examine them more fully in the next chapter.

In the remainder of this chapter we shall merely describe the historical process of economic and technological development in the analytical terms used so far, that is as a process of differentiation and integration of a societal function. Although Parsons only mentions the importance of modern science and technology in passing, I feel that its development warrants a separate discussion from the one on the development of the economy, for technological advance is so obviously the most important symptom of the greater generalised adaptive capacity of modern societies.

The structural differentiation of science and technology – the modern stage

Technology is a ubiquitous aspect of *all* human societies, but modern

---

* Parsons treats the differentiation of the economy and technology *together* as the differentiation into *one* primary subsystem, namely that of 'adaptation'.

societies are often called technological societies (or societies in the 'age of technology'), first because of the complexity and the superiority of their technologies, which have greatly enhanced the adaptive capacities of these societies, and secondly because of the primacy and the relative autonomy of their technologies *vis-à-vis* the other spheres of society.

The centrality of the desire to control and alter nature to suit material human wants and needs, and the ever-increasing ability to do so, indeed are the most outstanding features of modern technological societies. Indices of man's power to carry out his purposes with respect to the material environment in which he lives show near straight upward curves since the commencement of the modern period.[13] Thus life expectancies, diffusion speeds, world speed records, average calorie intake, cutting-tool efficiencies, population size and, regrettably, man's killing capacities, have all dramatically improved since the onset of the modern stage in the general evolution of societies.

For logistic reasons alone one may postulate the accelerating growth of technology in human societies. For, if one accepts the rather self-evident thesis that, first, technological growth occurs through inventions and diffusion of inventions, and that, secondly, every invention is a combination of two or more existing technological elements into something new, then it follows that the more existing technological elements there are, the more the possibilities for new inventions, and hence the faster will be the speed of technological growth.[14] In other words, technological growth can be expressed graphically as an exponential curve, geometrically related to the size of the technological base. Such, indeed, had been the case in the evolution of human societies until modern times. But since then the *rate of acceleration* itself has been accelerating.

It would be as absurd to suggest that the Middle Ages lacked sophisticated technology as it would be to ignore the steady flow of ingenious labour-saving devices which some historians claim was prompted by medieval man's desire to shorten working hours in order to devote more time to religious activities.[15] More absurd still would it be to disregard the immense contributions of the Chinese to the science and technology of those days. Indeed, most of the trump-card inventions which Europeans for a long time prided themselves in having fathered, have recently been proven to be Chinese bastards probably transmitted to us by European merchant travellers in the twelfth and thirteenth centuries, for example gunpowder, paper and printing, the collar harness, the wheelbarrow, the magnetic compass, the mechanical clock, water-wheels and water-mills, and last, but not least, several forerunners of the steam engine.

None the less, even those historians like Needham,[16] who have paid such conscientious tribute to the efficiency of the Chinese in applying knowledge of nature to useful purposes, readily admit that the meteoric ascent of technology since 1600 in Western Europe is a different cup of tea altogether, and one that can only be explained with reference to the birth of the modern sciences.

Modern science hinges upon the mathematisation of hypotheses about nature, and this took off in Western Europe only with Galileo.* That is why the onset of the modern period is usually dated at the year 1600 (Galileo lived from 1564 to 1642). In terms of our Parsonian action scheme, we may interpret Galileo's contribution as a fundamental reorientation toward 'ultimate reality' from which an entirely new cultural system issued. For underlying Galileo's scientific attitude was the conception of the world as an interplay of *calculable* forces and *measurable* bodies. Mechanical principles were seen to rule the movements of the planets, the changes on earth and the structure of the tiniest insect alike. Such conception of the world presented a paradigmatic shift away from the Aristotelian–Christian worldview which had dominated the Middle Ages, and in which view the structure of the universe was fixed and unitary in character, and was, moreover, of a profoundly divine order in which all bodies, heavenly and terrestrial had their fixed place in the hierarchy. The heavenly bodies were seen to be more 'perfect' than the terrestrial bodies, and changes undergone by bodies on earth (that is all the phenomena in our life) were held to be paralleled and controlled by movements in the heavens above. In such a worldview there could be no ultimate distinction between physical events, moral truths and spiritual experiences. The physical world was seen to have a moral plan, and therefore the authority of the Church was seen to prevail over natural knowledge. By contrast, with Galileo a differentiation between the spiritual universe and the world of natural phenomena was set in motion, and fully achieved a century later in Newton's work. This differentiation is commonly referred to as secularisation, a process which had its counterpart in the social structure with the complete separation between Church and State. The physical world from that time was seen as subjected to *objective* laws, to be discovered by logical deduction and careful empirical observation.

With the birth of the modern sciences technology received an entirely new impetus. For the making of new inventions became a *separately institutionalised procedure* and was pursued as an activity in its own right. Inventions were no longer stumbled upon, nor were they the lucky outcome of trial and error. They were not even always done with a view to serve some immediate practical interest. Rather, they were the result of purposeful experimenting in order to test logically deduced hypotheses about nature. Scholars and craftsmen mingled (this was in itself a uniquely European venture which prohibitive social structures elsewhere prevented) *in order* to discover the laws of nature. Technology became the tool of science. The new scientific attitude, having separated the physical from the moral world, could and would no longer rely on the Church's authority, nor anybody else's for that matter. *Nullius verba* ('on the words of no man') became its grounding principle, and as a corollary a premium was put on tangible evidence, on observation repeatable at will by other members

* The following discussion on modern science and its relationship with technology owes a great deal to the well-written text on that subject by Charles Singer, *A Short History of Science to the Nineteenth Century*.[17]

of the scientific community. It is in this context that the services of technology originally were most needed, because systematic observation and experiment were made possible by technology, and technology was in turn an offshoot of scientific needs. Optical instruments, such as the telescope, the microscope and various hyperbolic and parabolic lenses, as well as pendulum clocks were constructed to suit the experimenter's requirements.[18]

Not only were science and technology autonomous *vis-à-vis* Church and State, they also became differentiated from the societal community generally. Galileo's distinction between the 'primary' and 'secondary' qualities of physical bodies, with the former being the definitive (and hence scientifically interesting) properties, yet the latter being what ordinary man's experience of the world is made up of, meant that scientists were separated from their fellow men. The scientist would study wavelengths, where his fellow men would only see colours. Thus science and technology became specialised activities carried out by separate subcollectivities with their own ways of communication, their own symbols, their own processes of socialisation, and above all their own principles of ethical conduct.

The structural differentiation of the economy – the modern stage

As was the case with technology, the general evolutionary tendency of the economy is also towards increasing differentiation, and autonomy of its functional operation. Adaptive upgrading both in the execution of this function, and of the society as a whole, is the outcome of this differentiation process, because, says Parsons, differentiation of the economy leads to more efficient management as well as greater mobility of resources, human, physical and technical.

In the early stages of societal evolution, economic activity is so embedded in the social matrix of the kinship organisation that only narrowly defined ascriptive commitments may release the productive forces of a community. In primitive and, in varying degrees, in traditional peasant communities* both productive activities and distributive exchanges are largely determined by the social status of the members. Paraphrasing an old slogan, the anthropologist Raymond Firth once pinpointed the principle involved in primitive economies: 'From each according to his status obligations in the system, to each according to his rights in the system.'[19] The famous kula exchange of the Tobrianders in New Guinea,[20] and the Jajmani system of prescribed hereditary service relationships between members of different castes in traditional India,[21] are oft-quoted examples of such type of economies.

Not only are human resources thus kept in social bondage but so are the physical assets of the community: religious, social and political

---

* Throughout the history of general societal evolution, peasant communities have tended to remain outside the mainstream of the cultural and social developments of the civilisations to which they belonged, or were only very marginally involved in it. Therefore, at their own local levels of organisation, they have usually tended to retain many of the characteristics of kinship organisation typical of primitive communities.

orientations determine the range of uses that the land and all that is in (or on) it may be put to; some areas may be designated as sacred and hence are not to be productively exploited; in other cases certain crops may be the focus of sacred cults and hence grown to the exclusion of other crops (for example the Yam cults in New Guinea). By logical corollary, much of the productive and distributive phenomena cannot even be defined as purely 'economic' – they serve other functions (integrative, religious and political) as well. In his study of ceremonial exchange patterns Marcel Mauss stressed the 'total' character of these phenomena:

> These phenomena are at once legal, economic, religious, aesthetic ... and so on. They are legal in that they concern individual and collective rights, organised and diffuse morality.... They are at once political and domestic, being of interest both to classes and to clans and families. They are religious; they concern true religion, animism, magic, and diffuse religious mentality. They are economic, for the notions of value, utility, interest, luxury, wealth, acquisition, accumulation, consumption and liberal and sumptuous expenditure are all present.... Moreover, these institutions have an important aesthetic side.... Nothing in our opinion is more urgent or promising than research into 'total' social phenomena.[22]

This lack of social differentiation of the economic, or indeed of any other function, in primitive societies is the reason why anthropologists do study these societies in their totality rather than along the unidisciplinary lines true of their colleagues who study modern complex societies.

At subsequent stages of social evolution, the economic sphere although partially freed from ascriptive ties is too closely subjected to the political sphere. Resources are requisitioned by politico/religious or cultural elites, and economic management takes place under state administration. The problem of the religico/political administrations in the archaic civilisations, and even of the rationalised bureaucratic state apparatus of the great historic civilisations, such as Ancient Rome and China, was that the mobilisation and management of resources inescapably reflected and served the interests of the dominant social class from which the officials were recruited. This feature made these state bureaucracies inherently stagnant and unstable. Although Max Weber loudly praised the technical superiority of the rationalised bureaucracy over all prior forms of state organisation of the economy – because of its emphasis on procedural correctness, because of its control of the basis of specialised knowledge, because of the calculability of its operation, and so forth – none the less he was at pains to point out its limitations. These limitations were that however perfect a bureaucracy as instrument of organisation and management might be, it ceased to be so once it stepped outside its limits. And it was in its very nature to step outside its limits. For the very qualities which made it such a technically superior form of administration would make it transcend its own sphere of competence and throttle the freedom, the initiative and the enterprise of any and every group or section in the society: 'The bureaucratic order destroyed every economic as well as political initiative of its subjects.... The stifling of private economic initiative by the bureaucracy is nothing specific to antiquity. *Every* bureaucracy has the tendency to achieve the same results as it develops; our own is no exception.'[23] It is not surprising, therefore, that Max Weber, and his faithful disciple, Parsons, see modern socialist states as presenting a retrogressive step in evolution. Weber frequently compared the future in a socialist society with that under the stagnant, enslaving bureaucracies of Ancient Rome and Egypt.

According to Parsons, it is only with the advent of an *autonomous* principle that economic life can become fully differentiated from the social, the religious and the political spheres. This autonomous principle is expressed in a *universal* medium of exchange, namely *money in connection with markets*, and its adaptive advantage lies in that it allows for the most efficient mobilisation and management of resources, for it permits every type of good or service to be expressed in terms of every other type of good or service.

> It thus directs attention away from the more consummatory and, by and large, immediate significance of these objects toward their instrumental significance as potential means to further ends. Thus money becomes the great mediator of the instrumental use of goods and services... and thus this universal 'emancipates' resources from such ascriptive bonds as demands to give kinship expectations priority, to be loyal in highly specific senses to certain political groups, or to submit the details of daily life to the specific imperatives of religious sects.[24]*

Of course, both money and markets developed at much earlier stages of societal evolution. In the form of coinage, Weber notes, money first appeared in the seventh century B.C., and both its functions, namely that of prescribed means of payment and of medium of exchange (of which the former was historically the oldest) were found in almost all societies of the intermediate stage of evolution, with varying degrees of pervasiveness. Similarly, markets were not unknown in antiquity. What, however, distinguishes money and markets in the modern stage from all previous eras is the extent and the scope of both. As a *general* medium of exchange, and covering not only *goods* but also *labour* as well as *money* itself (for example in the form of interest) money did not become institutionalised until the late Middle Ages in Europe, and only then in Europe. Neither did labour nor capital markets develop in any noteworthy degree anywhere until that period.

With the full and universal establishment of money and markets, all resources – labour, land, technical skills and crafts, and indeed money assets† themselves – became 'freed' from all prior traditional restric-

* It seems to me that even Karl Marx could not have given a more complete and accurate description of 'commodity fetishism'. And it is really intriguing to note how the same process which for Marx epitomised human alienation, both Parsons and Weber appreciate as 'freedom'.

† This, in effect, meant that for the first time accumulated wealth could be transformed into productive capacity.

tions and able to move to *wherever their uses would be most profitable*, again in monetary terms. The latter part of this sentence gives away the underlying ethos, or cultural value orientation, of which the full use of money and markets was only a necessary physical expression. *Economic rationality*, that is the desire to maximise benefits over costs, became the most dominant and generalised motive in human behaviour after the onset of the modern stage. For Weber, and for Parsons, this principle of economic rationality is the definitive characteristic of *capitalism*, and *not* the expropriation of the surplus value of labour which is capitalism's more definitive characteristic according to Marx. Concretely, in Weber's terminology, 'A rational capitalistic establishment is one with capital accounting, that is, an establishment which determines its income yielding power by calculation according to the methods of modern book-keeping and the striking of a balance.'[25]

For such capitalism to become the dominant mode of economic organisation several conditions had to be met, of which the extension of money and markets in the sense described above was only one. New supportive changes in the normative order, that is in the legal system of society, were needed to sustain the function of money as a general means of exchange for all productive factors. The most important of these were the institutions of *private property* and *contract*. Weber once defined money as the father of private property: for the monetary principle to be an effective allocator of economic resources all physical means of production – tools, machines, as well as raw material and land – had to become disposable properties of owner agents, be these private individuals or economic establishments. The same holds true for the factor 'labour'. Individuals must be legally free to dispose of their own labour. Thus capitalism ended the slavery and the serfdom of all prior epochs. Unfortunately, as Weber himself notes, an important requisite for the smooth functioning of capitalism is that persons exist who are not only legally free to sell their labour, but who are also economically compelled to do so: 'Rational capitalistic calculation is possible only on the basis of free labor; only when in consequence of the existence of workers who in the formal sense voluntarily, but actually under the compulsion of the whip of hunger, offer themselves, the costs of products may be unambiguously determined by agreement in advance.'[26]

For the same reason that capitalistic accounting required a starving stratum of workers in order to enhance the calculability of the production process it also required *calculable* law. As we have seen, such calculable law was only a realistic possibility once the legal system had completely freed itself from any political or religious patronage.

Furthermore, the condition of free labour which 'voluntarily' joins other factors of production in the capitalistic enterprise presents an important and, in evolutionary terms, unique aspect of the process of economic differentiation: productive activities become separated from the household/residential unit in which they were previously located. As a consequence, status obligations which used to define productive roles within the household/residential unit come to be replaced by *contractual* obligations arising out of voluntary agreements between free agents.

Both the new normative complexes of private property and contract, as well as the institutionalisation of an independent formal legal system, require by logical necessity, a new definition of the societal community. The concept of citizenship, which permits the legal inclusion of all adult individuals in the society, presented such a new definition.

In our previous discussion of cultural evolution we already pointed to the reorientation towards ultimate reality and the developing cultural system of the Reformation, which could legitimise this new inclusion of all individuals in the societal community.

Cultural systems, however, do not only legitimise the *collective* order of society, they also provide the guidelines for the *individual's action in the external world.* Economic rationality, that is the desire to maximise economic benefits over costs, could never have become the more dominant motive in the determination of human behaviour had it not been but an integral part of a wider cultural value system, of which science and technology (as described earlier) was yet another expression. This wider cultural value system was geared towards *rational action* in the world, 'involving the explicit definition of goals and the increasingly precise calculation of the most effective means to achieve them, in contrast to action arising from habit or from traditionalism as principle'.[27] In other words, it involved utility rationality as opposed to value rationality. This value system in turn found its legitimising religious basis in the reorientation towards ultimate reality as formulated in the Reformation. As Weber said, 'Magic and religion are found everywhere, but a religious basis for the ordering of life which consistently followed out must lead to explicit rationalism is again peculiar to western civilisation alone.'[28]

We have already noted how the religious orientation of the Reformation had kept the notion of dualism between ultimate reality and human reality, whilst abandoning the idea of world rejection and religious activity as the only means to salvation. In fact, the key to the Reformation, and through it to the whole ethos of the modern era, was the rejection of the very idea of mediated salvation.[29] In radical revolt to the Church's monopolisation of the access route to 'God' and the 'hereafter', Protestantism (Calvinism in particular) emphasised the direct and immediate relationship between the individual and his 'transcendent god'. Abandoning both churches and monasteries as the sole places of, and prayers and other such devotional exercises as the sole expressions of, religious activity. Protestantism announced the whole of everyday life, everywhere and every minute of it, as a valid arena to work out God's 'command' and to serve his 'glory'. Thus Protestantism became a highly ascetic belief system. It asserted that only activity – hard work – could serve to increase the 'glory of God'. And to serve the 'glory of God' was man's reason for existence. Waste of time was the deadliest of sins. Not only was labour and hard work an approved ascetic technique, it was also a good defence against all those temptations which Puritanism united under the name of the 'unclean life'. Therefore, the maxim, 'He who will not work shall not eat', held for the wealthy as well as the poor. A rich man who did not work but lived off his money was considered sinful.[30] At the same time, out of disgust with the religious bartering to which much of Christian religious practice of the time had fallen to, Protestantism stoically embraced the concept of 'predestination': for everyone without exception, 'God's providence' had prepared a calling, a station in life in which he should labour and fulfil his duties, but not only was man's position *in* life prepared by the 'holy providence', so was his ultimate destiny in the life *hereafter*.

These two principles (of asceticism and predestination) were Calvin's legacy to his people: man is on earth to serve the 'glory of God', yet, however ascetic and ethical his conduct, he can never 'purchase' his salvation, for, doomed or chosen, man's fate was fixed before birth.[31]

This belief system would have been truly unbearable, and therefore ineffective as a practical guide to the moral behaviour of masses of Christians, had it not been for a third principle which – Weber notes – was not present in the original teachings of Calvin, but which was added on to the doctrine by his followers in their pastoral advice. This third principle held that although man could not actually influence his 'fate', he could at least come to know about it during his own lifetime. In other words, he could receive the *certainty* of 'God's grace', and so rid himself of the fear of damnation. This certainty of 'God's grace', to be sure, descended on the individual relative to the strength of *his own faith* in it, but this faith, in turn, originated in something visible and tangible, namely intensified ascetism: 'Works are not the cause but only the means of knowing one's state of Grace and even this only when they are performed solely for the glory of God.'[32] And, although the pursuit of wealth as an end in itself was condemned, the attainment of it as a fruit of labour in a calling was seen as a sign of 'God's grace'.

In this manner the Reformation threw the entire weight of its sanctions behind the principle of rationality: purposeful action employing means that were effective in promoting calculable goals would yield worldly success, and this worldly success was a sign of God's blessing.

Indirectly, the Protestant ethic also served economic development by promoting capital accumulation. For what could the successful Protestant do with his wealth other than to continue to invest it in his business, in his worldly enterprise? He could not relax in inactivity, enjoy the fruits of his labour, or squander his money on 'sins of the flesh', nor indeed could he give it away to the poor and idle, as being poor and idle was a sure sign of God's damnation. As Weber says, 'When the limitation of consumption is combined with the release of acquisitive activity, the inevitable practical result is obvious: accumulation of capital through ascetic compulsion to save. The restraints which were imposed upon the consumption of wealth naturally served to increase it by making possible productive investment of capital.'[33]

The differentiation of economy and technology combined: industrialisation – the adaptive upgrading of the modern stage

At the beginning of this section we described 'technology' as the behavioural and instrumental form of the adaptive function of society, and the economy as its organisational context. And although we discussed their respective processes of differentiation and development separately, we mainly did this for the sake of analytical clarity. Historically, certainly from the eighteenth century onwards (Weber puts it even earlier),[34] the entire process of economic and technological differentiation and development converged and combined to bring about together that system of production and organisation of economic life for which the adaptive upgrading of the modern stage in societal evolution is best known, namely *industrialisation*. For, in industrialisation, science and technology are applied to serve the criterion of economic rationality, that is to reduce cost per unit of output and thus to maximise profits.

The application of non-human sources of production (such as the steam-engine) made possible by developments in science and technology is a result of this merging of economic and technological developments. Small wonder that some scholars define 'modernisation' as hinging upon the application of non-human sources in production.[35] It is at any rate the most distinctive feature of industrialisation. The other characteristics of industrialisation follow more especially from the differentiation of the economy: specialisation and extensive division of labour, and the combination of the three factors of production – labour, raw materials and technical equipment – in special work locations such as factories away from household units. 'Industrialisation'

> is that system of production that has arisen from the steady development, study and use of scientific knowledge. It is based on the division of labour and on specialisation, and uses mechanical, chemical and power driven as well as intellectual aids in production. The primary objective of this method of organising economic life ... has been to reduce the real cost per unit of producing goods and services.... The resulting increases in output per man-hour have been so large as to stagger the imagination.[36]

Defined in this wider sense we can also include the commercialisation and the modernisation of agriculture into the definition of industrialisation. For modernisation of agriculture precisely involves the use of scientifically developed techniques to increase the fertility of the soil, to protect crops from insects, and to introduce higher-yielding varieties of seeds, as well as the use of power-driven aids in production. Commercialisation of agriculture means specialisation for markets, again with a view to maximise output and to reduce cost.

The final sentence of the definition points to the heart of the adaptive upgrading process that resulted from industrialisation as the combined outcome of economic and technological differentiation, namely a staggering increase in *productivity per man-hour*. The material benefits of this increased productivity have been visible for all to see, and have captured the imagination of the peoples in lesser-developed nations: rising *per capita* incomes; the ability to sustain larger populations; rising investment ratios yielding even faster-rising incomes and productivity; rising standards of living (at least material) for the masses of the population; and greater military power for the national community. The 64,000-dollar question is, of course, how far this increased productivity can be obtained without adopting the entire method of organisation of economic life along with its wider contextual social, cultural and political aspects. This problem will occupy us in the next chapter.

3

Neo-evolutionary Theory, Structural Functionalism and Modernisation Theories

One of the more conspicuous features of neo-evolutionism, the observant reader may have noticed, is the almost total lack of concern with causal explanations of social change. Indeed, as a proper theory of social change, neo-evolutionism has an appalling record, as many of its critics have, often angrily, pointed out.[1]

Neo-evolutionary theory does no more than present a *typology*, a classification of what it considers to be the major structural characteristics of societies at different stages of general (not historical or specific!) societal evolution. It does not attempt to say anything about the actual historic processes of transition from one stage to the next, apart from a vague reference to 'cultural diffusion' as the process whereby evolutionary breakthroughs are transmitted. How or why remains a mystery. Rarely, if ever, does neo-evolutionary theory even bother to speculate on the historical conditions that 'cause' the emergence of any one particular structural characteristic.

Parsons himself is well aware of this, but he does not see it as a shortcoming of his theory. He rather feels that contemporary sociological knowledge is not yet up to the task of 'dynamic' analysis. He claims that:

> Structural analysis must take certain priority over the analysis of process and change.... One need not develop a truly advanced analysis of the main processes of social change in order to make general claims about the *structural patterning* of evolutionary theory. This fact is well established in biology, where morphology, including comparative anatomy, is the 'backbone' of evolutionary theory.[2]

In other words, comparative analysis of the structural characteristics of different societies at different historical periods, as well as of different societies in the same historical period, juxtaposed against the evolutionary criterion of 'greater generalised adaptive capacity', may still yield a *sequential* order of structural types. By generalising the differences between those structural types, one may then legitimately formulate an evolutionary paradigm (Parsons prefers this term to the word 'theory'), for example evolution is a process of differentiation, adaptive upgrading, inclusion and value generalisation.

Whilst not criticising this methological perspective *per se*, I am in sympathy with those critics who accuse Parsons and other neo-evolutionists for sneakily bringing in a historical theory of change through the back door, because their choice of specific societies as illustrations of each stage of societal evolution (using Egypt and Mesopotamia, China and India, Rome, Greece and Israel, and Reformation Europe as examples) suggests quite deliberately a *historical* sequel which is never properly backed up by historical research.[3]

These criticisms aside, Parsons's neo-evolutionary theory, as observed before, is a recast of a strong European intellectual tradition into the mould of a contemporary sociological theory – again closely connected with Parsons's own earlier work – which had been going strong for several decades before his work on neo-evolutionism came out. In his own work on structural functionalism Parsons takes exactly the same position *vis-à-vis* the study of social change; a study of the structural characteristics of society must precede a study of change. So structural functionalism, too, has very little to offer by way of explanation of social change.

Amazingly, it is this very structural functionalism which in the 1950s and in the early 1960s parented the *modernisation* theories which are amongst the most, if not *the* most, popular and prolific theories about social change in contemporary developing countries. How very odd! How is it that a theory that has nothing to say about social change has become the dominant intellectual force behind theories of social change? Such exasperating somersaulting in contemporary sociology seems to require some investigation. But the task is not a simple one, for there are at least three factors which, in conjunction, are responsible for this strange methodological contradiction, namely (1) the methodological premises of structural functionalism; (2) the concept of cultural diffusion, and (3) the ethnocentrism of western social scientists, U.N. experts, World Bank advisors, and all those other westerners whose well-paid job it is to advise underdeveloped countries on how to become developed.

THE METHODOLOGICAL PREMISES OF STRUCTURAL FUNCTIONALISM

As a formal theory of society, structural functionalism[4] has four basic premises:

(*a*) Society is a system, that is a whole comprising interdependent parts;

(*b*) This systemic whole comes before the parts, meaning that one cannot understand any single part (that is cultural beliefs, legal institutions, social patterns of family organisation, political institutions, or economic/technological organisations) except by referring it to the larger systemic whole of which it forms a part;

(*c*) Understanding a part by referring it to the whole occurs by seeing that part as performing a function for the maintenance, the equilibrium of the whole. Thus the relationship between the parts and the whole is a functional relationship;

(*d*) The fourth premise is the most important for our purposes here:

it is really a logical derivative of the first and the third; the interdependence of the parts is itself a functional interdependence; the parts are mutually supportive of each other, mutually compatible with each other, and it is precisely in this mutual compatibility that they serve to maintain the whole. Parsons refers to this mutual compatibility of the components of the social system when he speaks of *structural imperatives*, or of the *imperatives of compatibility between structures* (that is economic, social, political and cultural). Imperatives of compatibility are 'those which limit the range of coexistence of structural elements in the same society, in such a way that, given one structural element such as a given class of occupational role system, the type of kinship system which goes with it must fall within certain specifiable limits'.[5]

In structural functionalism, the preoccupation with these imperatives of compatibility, or 'functional reciprocities' as they used to be called in the structural-functionalist School of anthropology,[6] has relegated the concern with cause and effect relationships to a secondary place. Thus, it became for structural functionalists a first priority, for example, to study, in our modern societies, the compatibility of:

(1) the full monetisation of the economy and the legal institutions of private property and contract;

(2) the industrial–occupational complex and the prevalence of nuclear and conjugal family patterns;

(3) the industrial–occupational complex and the relatively 'open' social structure (that is with its relatively high degree of social mobility);

(4) the pattern of extensive occupational specialisation and the formalisation of education, with the subsequent high literacy rates for the adult population;

(5) the differentiation of the main institutional spheres, the extensive occupational specialisation, and the structure of competitive pluralism; and

(6) the structure of competitive pluralism and the democratic organisation of the polity with its fully enfranchised citizenship and the multi-party system.

There was also seen the need to understand the compatibility between all these structural features simultaneously.

Understanding the 'compatibility' between the structures or components of society comes methodologically very close to what Max Weber used to call 'understanding at the level of meaning' as distinct from 'understanding at the level of causality'. But Weber did urge his fellow social scientists to study social phenomena at *both* levels, that of meaning and that of causality, and in his many careful historical works he certainly led the way.

However, in structural functionalism, the study of meaning – as it were – has obtained methodological priority, and it is because of this that structural functionalism has exerted such enormous influence on *modernisation theories*. For here we move one more and final step away from the study of causality when the application of the principle of structural compatibility comes to *replace* the dynamic analysis of social change altogether. Admittedly, the formal subject of modernisation theories has never been social change in general, but rather the processes of social change whereby less-developed societies become more developed. In other words, modernisation theories have always had a teleological concern with social change. As it happens, for less-developed societies the economic/technological complex of the developed societies constitutes the focal point of their interest in development. And for good reasons, it seems, since it is in this economic/technological complex that the superior adaptive capacity of the developed societies is demonstrated. Whatever 'image' developing countries may have of their own future, the undeniable fact is that they all, without exception, envisage economic development as the most important component of it. In other words, in these images of their future societies one structural component is 'given' or logically determined. And, because of the principle of structural compatibility, modernisation theories have tended simply to list the other structural features which are compatible with this economic/technological complex.

The reference to modernisation theories in this chapter, is a somewhat blanket reference to all those contributions in the field of *economic development and social change*, whether comprehensive or piecemeal, whether theoretical, empirical or policy orientated, which explicitly or implicitly have adopted the structural-functionalist premises in (1) viewing developing societies as *social systems* undergoing social change consequent upon the 'introduction', the 'impact' or the 'superimposition' of Western/technological institutions, and (2) in *preconceiving the direction* of this change on grounds of the principle of compatibility.

Of course there are differences amongst these approaches. A great many differences I would say: in focus, in scope, in the degree to which theoretical speculation is matched by empirical research, and even in the ideological concern with the outcome of the modernisation process.

Some theorists, like Smelser,[7] have focused more on the structural *discontinuities* of the modernisation process, for example on the problems that arise during the period of transition when old social orders are disintegrating and new 'compatible' ones are slow in being institutionalised. Other approaches have ethnocentrically advocated modernisation *policies*, for example the simultaneous transfer of western political, social and cultural structures together with the diffusion of the economic/technological complex, as a means to facilitate the transition. In this category belong particularly the experts of U.N. agencies and of the World Bank, whose task it has been not only to advise governments of underdeveloped countries, but also to secure the maximum economic success of their agencies' contributions in 'aid'.[8] In this class of advisors also belong the many western economists who have written textbooks on economic development, such as Rostow, Lewis, Hagen, Kuznets and Kindleberger.[9] In their efforts to transplant economic categories such as 'capital formation', 'labour', 'market mechanism', and so on they became aware of the intricate web of social and cultural institutions in which – in Western developed economies – the economic phenomena corresponding to these categories are embedded. Hence, in their economic-development theories, they recommended the reorganisation, that is westernisation, of social and cultural institutions as 'prerequisites' or as 'conditions' for economic growth. Again, others have concentrated on the aptitudes of the 'modernising elites' of underdeveloped countries in managing the development process, and some have gone as far as to organise training courses for economic elites so as to imbue them with the values and motivations appropriate to modern economic behaviour.[10] Some approaches have seen modernity as a downright wonderful thing, the crowning achievement of human evolution, and they have reckoned the underdeveloped countries to be lucky to be given the opportunity to catch up so quickly without having to go through the long and tormenting historical sequel of development in the West;[11] others still have bemoaned the disintegration of idyllic forms of traditional social solidarity and the disappearance of tribal social values.[12] However, despite all these differences, all modernisation theories have in common the belief that because of the cultural diffusion of Western economic/technological processes and because of the compatibility of societal structures, developing countries in the long run *inescapably* will come to take on the characteristics of the developed ones.

THE CULTURAL DIFFUSION OF THE ECONOMIC/ TECHNOLOGICAL COMPLEX: PATTERN VARIABLES

If, as is the case in modernisation theories, one sees that the impact of the diffused economic/technological complex in the long run is to generate *inescapably* a total transformation, that is a westernisation of the recipient countries, then methodologically this must imply that those who hold this view assume the economic/technological complex to *carry within itself* the reproductory cells of the remaining, non-economic yet compatible structural elements of the developed western societies. And although several generations of anthropologists had indeed implicitly taken this line, it was not until Parsons's work on pattern variables that this assumption was made methodologically explicit.

In his structural functionalist theory, long before he wrote his neo-evolutionary paradigm, Parsons had developed the concept of 'pattern variables'. Pattern variables are alternative patterns of value orientation in the role expectations of the actors in any social system. They are the points of interpenetration of the structural characteristics of the social system into the role expectation of the individual actor. In this way, pattern variables become the immediate determinants of social behaviour. The structural characteristics of society, in other words, are reflected in the role relationships of the individual actors in society.[13]

Parsons presents five contrasting pairs of value orientations which in their different combinations account for the nature of role relationships in any society, and hence typify the structure of any society. These five pairs of contrasting value orientations in role relationships are concerned with the following.

A. The *motivational basis* of action in the role relationship. Here Parsons distinguishes between those patterns of role expectations where role fulfilment is seen to be *affectively rewarding* or emotionally gratifying, and those where reward for role fulfilment is *affectively neutral* to the role incumbent. Thus, for example, when a child obeys its mother the reward it expects is perhaps a nod of approval, or a hug or kiss, all of which are affective rewards, but the wage or salary one gets for doing one's job are affectively neutral rewards.

The second motivational pair of alternative orientations concerns the question whether role obligations allow the role incumbent to pursue any private interests whilst carrying out the role, or whether the role obligations do not recognise any pursuits other than those in collective interests. These alternatives are called *self-orientation versus collectivity orientation.* For example, the subordination of a public servant's private interests in the carrying out of his public duties is an example of the collectivity orientation of the role. On the other hand, the role of entrepreneur implies the legitimate pursuit of private interests within the role context.

B. The *value standard* of the role content. This concerns the substantive definition of the role obligation. When the definition of the role obligation is couched in terms of universally valid precepts, which therefore transcend the limits of the particular role relationship, then we speak of a *universalistic* definition of the role obligation. When, on the other hand, the definition of the role obligation derives from an appreciative evaluation of the particular social object (for example the partner in the role relationship) we speak of a *particularistic* role definition. For example, role obligations such as those of economic agents in modern societies derive from general moral precepts such as 'contracts must be honoured', or 'maximise one's profits', or 'first come, first served', or 'putting the right man in the right place'. These are universalistic definitions. On the other hand, role descriptions in such terms as 'I must try to help him because he is my tribesman', or 'I must give him some more credit because he is my cousin', are particularistic role definitions, because the role activity is confined to a particular social object or class of social objects.

C. The *evaluation* of the role incumbent. When the focus is on the role incumbent's performance in his role, we say that the basis for evaluation is that of *achievement.* When, on the other hand, orientation to the actor's social and physical attributes predominates, the basis for evaluation is *ascriptive.* In the former case, it is important what a role incumbent does or can do, for example how many qualifications, degrees, experience, and so on, he has; in the latter case it is more important who a person is, for example from what family background he comes, whether he is male, or female, rich or poor, and so on.

D. Finally, the *scope* of the role relationship presents the last pair of alternative patterns. A relationship between the participants in the role relationship may be *functionally specific* or *functionally diffuse.* An example of the former is that between employers and employees in modern-type industries, and an example of the latter is that of a patron and his apprentice in medieval craft guilds, where the patron was not just employer, but also teacher and guardian to the apprentice, who in turn was not merely an employee, but also pupil and adopted son to the patron.

The suggestion that each and every combination of these five times two pattern variables yields a different type of role relationship, and hence different types of social structure, would lead one to suppose that there are 2^5 (equals 32) possible types of social structure. However, because of the structural compatibility between a number of these pattern variables (especially *B*, *C* and *D*) the number of empirically likely combinations is reduced.

Thus Parsons claims that the modern economic/technological complex with its high degree of occupational specialisation, and the application of the principle of economic and instrumental rationality, favours roles that are *functionally specific*, *achievement orientated*, *universalistic and affectively neutral.* Obversely, in the relatively undifferentiated structures of primitive societies and traditional peasant communities roles are likely to be functionally diffuse, ascriptive, particularistic and affectively rewarding.[14]

The methodological argument of modernisation theories of social change in contemporary developing societies seems to be that these internally consistent action patterns are built into the modern economic/technological institutions which underdeveloped countries so eagerly adopt (or which had been imposed upon them during the colonial period) and that these action patterns in turn send their reverberations through the entire social, cultural and political structure of developing societies.

Let us briefly follow these action patterns as they are thought to steam-roller their way through the wider societal structures.

(1) First of all the *universalistic* application of the norm of profitability to all economic resources – land, labour and capital – means that all these resources must constantly shift from lesser to more productive uses. The wider structural consequence of this is *mobility.* As Wilbert Moore once succinctly expressed it, 'If one were to attempt a one-word summary of the institutional requirement of economic development that word would be "mobility". Property rights, consumer goods and labourers must be freed from traditional bonds and restraints.'[15] The effects of the mobility of labour, typically, is *urbanisation.* Labour which is no longer productive in agriculture must be freed from the land and be grouped together in factories using concentrations of capital equipment and raw materials. Urbanisation then logically follows as a second-order consequence of the concentration of labour in factories as well as of the concentration of factories which occurs because of the cost-reducing potential of the proximity of industries.

(2) The geographical mobility of labour in conjunction with the *functional specificity* of modern economic roles (that is, the extreme division of labour coupled with the strict segregation of workplace

from household) in turn 'breaks up' traditional kinship patterns of family organisation and presses for *conjugal-type family patterns*, that is few kinship ties with distant relatives and an emphasis on the nuclear family unit of husband and wife relationships with their children.[16] This again, requires a formidable reorientation of these husband–wife relationships, with the *romantic-love complex* coming in as a necessary adhesive for the strong affectionate ties between the spouses, because these ties now have to take the brunt of the individual's need for emotive support.[17]

Nuclear family patterns are obviously not enough to prevent the individual's personal disorientation in the fragmented and impersonal life of town and industry. The disintegrating solidarity groupings of village and wider kinship communities gradually come to be replaced in the towns by new 'functional' groupings such as churches, halls, sports clubs, trade unions, and so on. These *voluntary associations* are typically modern integrative structures bridging the emotional gap between the tight but too small conjugal family group on the one hand and the anomic, fragmented modern urban society on the other.[18] And finally, it is often argued, the processes of urbanisation and industrialisation require a new solidarity at a national level. Uprooted from traditional systems of loyalty and security, and lost in a labyrinth of cross-cutting as well as competing economic interests, the individual is in need of a new identity and a new sense of belonging. *Nationalism* presents a unifying force that on the one hand may psychologically rationalise present dissatisfactions, for example by using former colonial powers as scapegoats, and on the other hand excite people to even greater efforts and sacrifices.[19]

(3) Not only is geographical mobility of the factor 'labour', with all its wider social, cultural and political ramifications, an institutional implication of the modern economic/technological structure, but so is *social mobility*. The application of the principle of economic rationality implies that people acquire status in a society as a result of their personal *achievements* rather than as a result of ascription by birth: 'The significance of this principle is obvious: just as the industrial system tries to combine non-human factors of production efficiently, so it will seek to maximise its return from wages and salaries by putting the right man in the right place. . . . Men are chosen not for whom they are and whom they know, but for what they can do.'[20] A universal consequence of industrialisation therefore is said to be the alteration of the principle of stratification in a society from ascription to achievement and the consequent social mobility of the individuals in that society. A contextual development is the need for *formal education* and vocational training to promote and standardise the levels of achievement, and so facilitate the selection and the recruitment of the human talents and skills needed in industry.

(4) The mobility of the other two factors of production, capital and land, hinges upon the institutionalisation of a new norm, the right to buy and sell property, be this a right on the part of private individuals or on the part of the State.[21] The *transferability of property rights* unavoidably disrupts communal ownership of land, as well as feudal

forms of ownership and tenure. The easy transfer of land to more efficient producers coupled with the commercialisation of agriculture (as yet another effect of economic rationalisation) results in the re-organisation of land into more efficient and larger holdings specialising in the production of cash crops, and this in turn tends to transform the social structure of the countryside, with a large section of the peasantry, for example tenant farmers and sharecroppers, as well as small independent producers, turning into a landless rural proletariat whilst the middle and rich peasants are in a position to benefit from the economies of scale and the application of modern technology.[22] Furthermore, the modernisation model assumes the absorption of this proletariat into the secondary and tertiary sectors of the industrialising economy.

Generally, modernisation theories have assumed that the institutionalisation of private property rights must also include private ownership of the means of production (that is of capital and technology as well as of land), and therefore they are theories of *capitalist development*. A necessary outcome of capitalist development is an over-all structural division of society into *classes*, namely into those who own the means of production and those who only have their labour to sell. Following Weber (and Parsons of course) who did not see the capitalist/worker dichotomy as the one and only definitive structural division in society, modernisation theories have tended to speak in terms of an increasingly *plural social order* characterised by not just two but many different *strata*. Modernisation theories have usually been very scathing about possible alternatives to capitalist development, such as socialisation of the means of production by the State – which I would refer to as State Capitalism – or other socialist approaches which do not only socialise the means of production but which also attempt to break away from the canons of economic rationality. We shall come back to these possible 'alternatives' of development in the last chapter of the book.

(5) The *universalistic* application of the principle of profitability in modern production organisations implies that all decisions as to what shall be produced, when, how, where, and by whom, should be made entirely on grounds of expectations as to what is going to yield the highest economic returns, and should not include any 'particularistic' observations. Thus, the economic decision to build a railroad, and the corresponding decisions as to where to build stations for regular stops, and so on; the choice of a site for a steel plant, or the choice of production techniques within a factory, all must be made entirely on grounds of their expected productivity, and not on the basis of the particularistic attributes of the economic objects, for example their familial, tribal, regional or religious background. The organisational context in which this principle of economic rationality is most easily universalistically applied, is that of the *rational bureaucracy*, for this, as we have seen already, is based on precisely a strict adherence to formal rules and on the rigid separation between personal and official life (for example the work role). That is why the rational bureaucratic organisation of the modern business corporation is thought to be

excellently suited to organise modern economic productive activities.*

(6) A logical corollary of universalistic role descriptions is that role relationships in modern economic settings are based upon *contract*, that is the contract defines the terms and conditions of the work role regardless of the role incumbent. However, the contractual nature of the role relationship in modern economic structures has its drawbacks, and therefore calls for wider supportive changes in, especially, the political structure of society. By making existing social relations irrelevant for economic pursuits it makes for a great deal of uncertainty. For example, if one enters into a business contract with someone whom one does not have any prior social relations with, and whom one may recognise to be of different kin, clan, tribe, or religious creed, and yet whom one – on purely economic grounds – has chosen as a partner to do business with, how then can one make sure that the person will honour his part of the contract and fulfil his side of the agreement? In pre-industrial societies, economic transactions are so enmeshed in social relationships that the question of lack of trust and confidence hardly arises. Firth writes that

> The (primitive) system operates because in addition to the immediate social satisfaction gained, some material reward is often eventually obtained. Concretely, the work which A does for B as a social obligation, is paid back in the long run by gifts which C makes to D. Here, B may be the father-in-law of A, D the son of A and C the brother-in-law of A, and D the son of A. The recognition of the long-term trend towards equivalence of service is one of the most important incentives to work in a primitive or peasant society.[24]

In modern economic production organisations and commercial transactions, on the other hand, where one is dealing with persons unknown, there needs to be a transcending body of authority and control, a 'third' party, that can uphold the new contractual social order and *secure the continuity of economic relationships.* In short, modern economic activity requires the formation of a *centralised territorial State.* But in order for this State to be effective, it needs to be represented by a legitimate national government, that is one recognised by all economic parties concerned in the new social order. And therefore, as American scholars are especially fond of pointing out, it must be a national government legitimised by a *democratic polity*, that is one where consensus in execution of leadership is mediated through the secret ballot-box by a fully enfranchised adult citizenry, and where the articulation of political interests occurs through a multi-party system which reflects the plural character of the new social order.[25] A further contextual development of political pluralisation is the institutionalisation of 'freedom of speech' and the articulation and education of

* Although Max Weber regarded the rational bureaucracy as the perfect organisational instrument, he was frantic lest this perfect organisation should become the 'only' (that is *public*) organisation of economic life. Therefore, Weber wanted bureaucratic organisations of *private* capitalism to compete with the bureaucracy of government and state administration, so that they would counterbalance each other and hold each other in check.[23]

public opinion in an independent system of communication, *the free press. Literacy* of the adult population, again, is a contextual requirement of the proper functioning of both the free press and of public participation in the polity.[26]

(7) Last but not least, the *affectively neutral* forms of gratification operative in modern fully monetised economic systems require the expansion of the consumer market, because wages and the things one can buy with them provide the only form of motivation to work in the modern economic system. The transitional difficulties of this pattern variable were clearly brought out in the backward-sloping supply curves of labour in the early industrialisation periods of European overseas colonies. Great was the confusion of colonial entrepreneurs when native workers appeared unresponsive to wage incentives. Natives did not seem to want to work harder when offered more money; in fact, maddeningly, the more money offered the less they would work.[27] In Europe in an earlier century this 'irrational' and 'uneconomic' behaviour had prompted the adoption of mercantilist policies which stated that, 'therefore', wages should be kept as low as possible, thus preventing 'idleness, drunkenness, gambling and immorality' and so 'forcing the worker into virtue'.[28] The same solution was applied in the colonies. Anthropologists, though, pointed to the limited consumption ceiling of the native worker, who much like his European colleague of an earlier century, did not yet have the capacity to consume *ad infinitum*, both psychologically from a point of view of his wants and needs, as well as socio-economically from a point of view of the consumer goods available to him.[29] The effectiveness of affectively neutral, financial incentives is therefore related to the expansion of the consumer market and to the cultural reorientation of recipients towards the goods and services offered by the industrial system of production.

EUROPEAN ETHNOCENTRISM: MODERNISATION THEORIES AND DEVELOPMENT POLICIES

In this chapter we have taken a bird's eye's view of one very popular kind of theorising about development in underdeveloped countries, and we have noted how this theorising has its roots in a long-standing tradition of Western reflection on the structure and the evolution of modern Western societies.

In the opinion of the author the continuing influence of this theorising upon development *policies* can hardly be exaggerated. And, as suggested earlier, the key to this influence lies in the confusion of the methodological construct of 'structural compatibility' with that of 'causes' of social change. Because modernisation theories have viewed the total transformation, that is westernisation, of developing countries to be an inescapable outcome of successful diffusion of the Western economic/technological complex, by methodological *reversal* it is argued that a reorganisation of existing social and cultural as well as political patterns in anticipation of their compatibility with the diffused Western economic/technological complex may in fact *facilitate* the very process of this diffusion itself.

This monumental theoretical error – which to be fair was not always committed by the theorists themselves – has in fact been made and continues to be made by modernisation policy-makers such as those employed by Western governments, U.N. organisations, the World Bank, and so forth. Thus the list of features shown in the table below has become ever so many 'indicators' of social, political and

Indicators of development

degree of urbanisation
literacy rates and vocational training
newspaper circulation
political democracy (as measured by the existence of a multi-party system and the regular executive transfer through secret ballot elections)
free enterprise
secularisation (that is institutionalisation of 'rationality' as the dominant behavioural norm)
degree of social mobility
occupational differentiation
proliferation of voluntary associations including, for example, trade unions
national unity (as opposed to ethnic and denominational factionalism)
nuclear family patterns
independent judiciaries

cultural development, and from that position has frequently been promoted to 'conditions' for economic development.[30] This in turn has had immediate consequences for policy-making:

(*a*) Aid flows were and still are frequently directed to such supportive changes in the socio-political structure, and in doing so they have often created more problems than they have solved;

(*b*) Industrialisation, narrowly defined as a transition from agricultural to industrial activity, was embraced as a hallmark of development, and crash programmes both of industrialisation and of agricultural modernisation set adrift a rural proletariat which flocked to towns causing over-urbanisation rather than urbanisation;[31]

(*c*) Much money went into formal education for people for whom there were no jobs;[32]

(*d*) World Bank loans were tied to the pursuit of monetary policies which favoured 'free enterprise' and the emergence of a bourgeois capitalist class,[33] and American business schools designed appropriate curricula to train these bourgeois capitalists to become a proper managerial and entrepreneurial elite;[34]

(*e*) Western aid programmes were strictly confined to those countries where the existence of parliamentary democracy (or a resemblance of it) conformed to the 'Western model', whilst American covert C.I.A. activities were directed to toppling 'undemocratic regimes'. The manifest justification for such American intervention was time and again the 'protection' of democracy and the rights of 'opposition' in these countries, and in their (that is these countries') very own best interests.[35]

If it had not been for the scholarly proliferation of *bona fide* structural-functionalist theories of modernisation, students of development would have more readily seen through the true nature of the Western philosophy of development. For structural-functionalist theories of modernisation have in fact very usefully served as an ideological mask camouflaging the imperialist nature of western capitalism. The western capitalist system for its very own survival needed and still needs to *expand*. It had and has to spread its tentacles further and further across the globe. But the success of its expansion depends largely on its ability to reproduce, wherever it reaches out to, the structural conditions under which it operates at home. Thus westernisation becomes a tool of imperialism. This theme, which is the calling tune of the *neo-Marxist* approaches to development, we shall be discussing in the next part of the book. The methodological merit of the neo-Marxist perspective is that it does not assume developed or underdeveloped societies to be 'self-sufficient' social systems, but rather views them as interconnected parts of an over-all global social system. Its second methodological merit is that it studies historical *causes of* social change in the underdeveloped world; it does not mistake or confuse structural compatibility with causes or consequences of change, and hence it is not misled into believing that modernisation is the inescapable outcome of the diffusion of Western/technological processes. Rather it sees the continued underdevelopment as the inescapable outcome of this diffusion process.

Modernisation is tantamount to 'independent development'. But, as we shall see, the very process of so-called diffusion creates dependence and hence negates the possibility of modernisation.

Part Two
Development as Interaction

4

The Development of Underdevelopment: Mercantilism, Colonialism and Neo-colonialism

> Nearly all our major problems have grown up during British rule and as a direct result of British Policy; the princes; the minority problems; various vested interests, foreign and Indian; the lack of industry and the neglect of agriculture; the extreme backwardness in the social services; and, above all, the tragic poverty of the people.
>
> Nehru: *The Discovery of India*, 1946.

In Part One of this book we studied processes of societal evolution and change on a very abstract plane. From a standpoint of pure sociological theory, societal evolution presents a process of increasing structural differentiation of the functions performed by society. Thus, from a sociological perspective, structurally complex societies are more developed than structurally simple societies.

In presenting such a paradigm for the study of societal evolution society was treated as if it were a social system in isolation; as if it were existing in a vacuum unrelated to other societies; and, if contact between societies occurred at all, it was conceptualised as 'cultural diffusion' – that gentle transmitter of evolutionary universals. Evolutionary theory never considered the possibility of structural relations between societies.

Historical societies, however, do not fit this abstract picture. Not only do we witness frequent culture contact and economic exchanges between societies, but history reveals that this interchange frequently leads to the political, economic and cultural domination of one society or group of societies over the other. The upshot of such domination, sociologically speaking, is a form of *inter-societal stratification*, often prompting the conquerors toward greater complexity whilst reducing the conquered societies to more simple forms of societal organisation.

The historical reality of inter-societal stratification does not impair the relevance of an abstract paradigm of societal evolution as presented in previous chapters. Rather, such a paradigm can assist us in analysing how external penetration arrests the internal process of societal evolution and in so doing creates 'problems of development', or 'symptoms of underdevelopment'.

Turning our attention directly to recent periods of history, we observe that it has been the very expansion and development of one group of *nations* (Western Europe) which has disturbed and prevented the process of indigenous societal evolution of another group of *nations* – and, indeed, continues to do so up to the present time. Expressed differently, one may characterise contemporary developing societies as societies whose internal developments have been disrupted as a result of the impact of the West. It is precisely at the crossroads of the process of internal societal evolution on the one hand, and the penetration of Western domination on the other, that we can identify and understand some of the problems characteristic of contemporary developing societies. The metaphor 'crossroads', however, seems helpful only at a superficial glance. From the point of view of the resulting problems of development today, the impact of the West upon these unfortunate territories clearly falls into not one but two distinct periods and patterns: the mercantilist and the colonial.

Western supremacy in world trade during the mercantile period (roughly between 1500 and 1800) and its subsequent economic exploitation of the African, Asian and Latin American continents, whilst helping to accelerate the social evolution of the societies in Western Europe, had an arresting and frequently degenerating effect upon the social evolution of the societies in the three other continents. This argument, which is drawn from Marxist theories of the development of underdevelopment, will be presented in this chapter.

These centuries of economic exploitation resulted in a global stratification – or polarisation – of 'advanced' and 'backward' societies, but this very polarisation presented a historical paradox; the resulting disparity in societal evolution between the trading partners turned into a bottleneck for the further advancement of the West. Such a situation invited a period of colonialism; a period of subjugation by integration, and integration by subjugation. The economic production of the underdeveloped regions could only continue to serve the needs of the advanced areas if the producers were to adapt to, and therefore adopt, Western patterns of culture and technology as well as Western forms of social and political organisation: they needed to become 'uplifted' to a level of societal organisation closer to that of the advanced world. At the same time the imperialist rivalries between the advanced states meant that only direct political domination and administration would ensure the effective incorporation of the underdeveloped economies into that of the 'mother country'. The superimposition of and hence the attempt to 'diffuse' Western forms of social, political and economic organisation characterised as they were by a high degree of structural complexity, on the by that time retrogressed forms of economic, political and social organisation of the 'backward regions', created exactly those classic problems of *tradition versus modernity*, of *dualism*, and of *discontinuities* that have captured the attention of 'bourgeois'-functionalist sociologists and anthropologists for decades. These typical problems arising out of the contact between 'advanced' and 'backward' cultures and societies we shall discuss in Chapters 5 and 6.

It is important now to note that Marxist and bourgeois-functionalist perspectives on problems of contemporary developing societies are not necessarily incompatible. Gunder Frank, who has perhaps most systematically reviewed the theoretical premises of the functionalist, diffusionist and dualist approaches to the 'sociology of development', criticises them for 'attributing separate and largely independent economic and social structures to the developed and underdeveloped sectors, each with its own separate history and dynamic, if any'.[1] He argues that the supposed structural quality is contrary to both historical and contemporary reality, and he pleads for a sociology of development anchored in structural and developmental holism.[2]

It is the contention of this book that the two approaches *are* compatible, and that 'structural and developmental holism' is attained by treating them as complementary. The difference in structural complexity between the 'modern' and the 'traditional', and the problems created by their interaction, are empirically identifiable social situations in contemporary developing societies, calling for a structural-functional analysis. On the other hand, it is the historical Marxist approach which is theoretically better equipped to explain the dialectical process of international development that culminated in these very structural discontinuities of contemporary developing societies.

DEVELOPMENT AND UNDERDEVELOPMENT; A DIALECTICAL PROCESS

The thesis that the underdevelopment of today's Third World resulted from its being brought into the orbit of the capitalist expansion of the West can be traced back to Marx's discussion of foreign trade and the expansion of capitalism.[3] Since then, the theory has been variously elaborated by many scholars, the better known amongst whom are Lenin,[4] Paul Baran,[5] and André Gunder Frank.[6] Since a proper review of each of these theories falls outside the scope of this book, it will suffice to present the basic tenets of the thesis.

Fundamental to the theory is the conception of a *dialectical* relationship between the development of the First World and the underdevelopment of the Third World. The term 'dialectic' refers to a two-way causal connection. What is implied is that the West developed precisely because it was underdeveloping the Third World, whilst the Third World became underdeveloped in aiding the ascendancy of the West. Historically this dialectical relationship has unfolded in three distinct stages: a mercantilist-capitalist stage, a colonial stage and a neo-colonial stage.

MERCHANT CAPITALISM

Roughly from about the middle of the sixteenth century to the end of the nineteenth century, European traders combed the coasts of Africa, Asia and Latin America in search of slaves, spices and gold, and in conquest of existing trade routes. Ironically, as Keith Griffin has pointed out,

> it was the combination of Europe's military superiority and her relative material poverty which shaped events in the early phase of

> European expansion. Western ascendancy was made possible by advanced military technology and it was made necessary by the inability of Europe to engage in trade on equal terms with the wealthy nations of the East. Asia had much that Europe wanted – but Europe could offer almost nothing that was desired in Asia.[7]

Said the Chinese Emperor scornfully to King George III: 'Our celestial empire possesses all things in prolific abundance', thus indicating that China had little need for English imports.[8] But the superiority of their gunned ships – developed in the course of the fourteenth and fifteenth centuries[9] – soon allowed the European traders to subjugate their overseas trading partners to a pattern of commerce hardly distinguishable from outright plunder. 'For the purpose of buying goods at the cheapest price possible an ultimate political control over the countries they traded with was a *sine qua non* for the policy of the merchant company.'[10] Thus, in the seventeenth and eighteenth centuries foreign commerce in Asia and Africa went hand in hand with political concessions to monopolistic corporate bodies of merchants, namely the East India Trading Company, the Levant Company, the Africa Company, the Vereenigde Oost-Indische Compagnie, and so on, whilst in Latin America territorial conquest and consequent pillage resulted at the outset from individual action under state patronage. The treasures thus 'captured outside Europe by undisguised looting, enslavement and murder flowed back to the mother country and transformed into capital'.[11]

This, then, is the first important tenet of the development of underdevelopment thesis, namely that the unilateral transfers of wealth from the African, Asian and South American continents constituted, in historical terms, an almost instantaneous increment in the economic surplus of the European nation states, a surplus which largely concentrated in the hands of capitalists who could use it for purposes of industrial investment.[12] Few economists today would deny that the successful transformation of an agricultural into an industrial society requires a substantial increase in capital accumulation. Some confidently place the savings–investment ratio required for industrial take-off at 10 per cent of the gross national product.[13] The development of underdevelopment thesis maintains precisely that overseas trade in the seventeenth and eighteenth centuries constituted an exogenous contribution to Western Europe's capital accumulation, and hence presented a necessary, though not sufficient, cause of Europe's economic development. Note the distinction between necessary and sufficient causes! Mercantile capital did not always find its way into manufacturing enterprises and thus help finance industrial expansion as it did in England in the seventeenth and eighteenth centuries. Much depended also on the presence of favourable socio-cultural conditions, such as (1) the existence of a traditionally respected class of artisans and craftsmen ready to become the new captains of industry, as well as a class of men who had been expelled from the land by an agricultural revolution and who were searching for work in the towns; (2) a scientific revolution which, after all, brought the steam engine; and (3) a

Protestant Ethic which mitigated against extravagance and admonished even the wealthy to work. In the absence of these favourable socio-cultural factors, the abundant flow of mercantile money led to the wasteful life of its possessors and the ultimate decline of the nation, as was the case with Portugal and Spain.[14] Indeed, in the eighteenth century the Spanish and Portuguese empires became most profitable markets for English manufactures. In these two 'metropolis' countries, commercial prosperity consequent upon the South American conquest led rather to a conscious abandonment of a policy of home industrial expansion.[15]

One is reluctant to make the relationship between the English overseas adventures and the Industrial Revolution as closely linked as some authors do.[16] Baran quotes Brook Adams: 'Very soon after Plassey (1757) the Bengal Plunder began to arrive in London and the effect appears to have been instantaneous: for all the authorities agree that "the industrial revolution", the event which has divided the nineteenth century from all antecedent time began with the year 1760.'[17] This indicates that there is little doubt that the accumulation of merchant capital did indeed help finance the Industrial Revolution.

Whilst economic and hence social ascendancy in the West was rooted in a pillage of overseas economic surpluses, the systematic loss of these surpluses at the same time removed the opportunity for economic advancement in the territories where the West 'traded', and so arrested their further internal development. This is the second basic tenet of the development of underdevelopment thesis. And not only was their internal development discontinued, their confrontation with the West actually had a *regressive* effect on the level of societal evolution which they had already reached. This regression in social structural complexity was the combined outcome of *demographic*, *economic* and *political* decline.

Demographic decline

Whereas in Africa the slave trade accounted for a gruesome loss of manpower in number well above that of the estimated ten to fifteen million slaves who arrived in the Americas,[18] on the American continent it was not so much the *trade* in slaves as 'the combination of brutality, slaughter, high tribute, slavery, forced labour for goldmining, destruction of the social framework, malnutrition, disease and suicide' which wiped out large sections of indigenous peoples.[19] Colin Clark estimates that in Latin America as a whole the population declined from 40 million in 1500 to 12 million in 1650.[20] Elsewhere, in Asia, the introduction of European diseases decimated indigenous populations.[21] Since social differentiation, and hence a peoples' degree of structural complexity, is in part a function of the division of labour consequent upon population growth and population density,[22] one can well imagine the socially debilitating effects of depopulation.

Economic decline

Economies degenerated as assistance in European trade became a dominant feature. In Africa the slave trade replaced much of the traditional trade, and the pre-eminence of trade itself prevented or reduced diversification of economic activity: 'Even the busiest African in West, Central or East Africa was concerned more with trade than with production, because of the nature of the contacts with Europe, and that situation was not conducive to the introduction of technological advances.'[23] The goods received in exchange for the slaves reduced existing levels of economic activity still further: fire-arms helped dislocate whole societies, and textiles and metalware undermined Africa's own industrial production. The relinquishing of certain productive activities in the face of European competition even occasionally led to technological regression: the abandonment of traditional iron-smelting in most parts of Africa is a case in point.[24] Of course, economic regression resulting in social de-differentiation was even more dramatic on the Asian subcontinent. The industrial decline of India in the pre- and early-colonial period has been too well documented by others to need any further discussion here.[25] Emphasis in India, as elsewhere, from then on was placed on the production of agricultural cash crops to suit the needs of industrial expansion in Europe. Marx, writing on the cotton trade, observed that 'A new and international division of labour, a division of labour suited to the requirements of the chief centres of modern industry, springs up and converts one part of the globe into a chiefly agricultural field of production for supplying the other part which remains a chiefly industrial field.'[26]

Even if the period of merchant capitalism prepared the way for this international division of labour, one should be cautious not to exaggerate the extent to which, in this period, European manufactures 'crushed' local industries, handicrafts and arts, and 'forced' a class of artisans to return to the soil – as is so often argued. For it was not until after 1815 that prices of European (especially British) manufactures 'began to fall dramatically both as a result of increasing mechanization and cost reduction in the factories, and in response to a sharp decline in the prime cost of raw materials and import duties into the United Kingdom'.[27] Platt argues that before 1815 local manufacturing in the South Americas, as well as in India and China, kept up the competition and quite successfully so.[28] Indeed one might add as a curious detail that African local textile industries initially suffered not only from the competition of European manufactures but also from the competition of *Indian* cloth which the European traders brought.[29]

Political decline

In the mercantile period, finally, the impact of Western commerce, conquest and gun-boat diplomacy eroded the political integrity and cohesiveness of the inflicted peoples. In Africa, where direct contact with the European until the nineteenth century remained an exclusive 'prerogative' of the coastal peoples, the debilitating effects of the slave trade, which encouraged inter-tribal warfare on a hitherto unknown scale, effectuated political decay. Political disintegration completed the process of social structural recession. No one needs reminding of the fate of the great pre-Columbian civilisations in South America, the crumbling of the Mogul Empire in India, the slow erosion of the Ch'ing Dynasty in China or the disappearance of the kingdoms of Java and Sumatra. Less widely known or publicised has been the political decay of African empires such as Songhai, Kanem-Bornu and the Hausa City states in West Africa, the Ethiopian and the Nubian states in the North-east, and the great Zimbabwe culture in the East.[30] One may in fact maintain that everywhere in the world until the European overseas expansion, politically centralised states could be found which encompassed social and economic structures of a degree of complexity and advancement comparable, and favourably so, with feudal Europe on the eve of its scientific and industrial revolution.

COLONIALISM

If the search for foreign produce initiated the era of European merchant trade, the desire for controlled market outlets for European manufactures was a dominant feature of the period of colonialism, the second historical phase in the dialectic of development and underdevelopment. Characteristic of this period was that increased territorial responsibility, due to conquests, gradually and at first reluctantly, replaced the comparatively arbitrary exploits of commercial plunderers with a more deliberate administrative system designed to adapt the socio-economic organisation of the colony to the needs of the mother country. In so doing, the Marxist view holds, European imperialism further underdeveloped these already disadvantaged territories.* Such a view, however, is only partially correct. In order to adapt the underdeveloped territories to suit the requirements of the by then highly advanced mother countries, the West at the same time had to impart its own technology, its methods of production, and hence its forms of social organisation, its cultural ideas and its political and legal structures. In short the West became the 'civilising agent'. The impact of this diffusion tended to concentrate in geographical locations near European settlements, thus making 'duality' the most outstanding feature of the colonial and, indeed, of the post-colonial society. We shall discuss this and other such features in Chapter 5 and 6.

Yet at the same time, and viewed from a historical perspective, the integration of the colonial economy as an appendage the mother country *can* be regarded as a process of further underdevelopment.

* Latin America, though formally in this period independent, for all practical purposes can be included amongst the colonies of Western Europe: 'South America . . . is so dependent financially on London that it ought to be described as almost a British commercial colony', says Lenin, quoting Schulze-Gaevernitz, and coins the term 'semi-colonies' to describe Latin America, as well as for example such places as China.[31] Indeed, from the point of view of the relationship between capitalism and underdevelopment, as set out in this chapter, Latin America in the nineteenth century, after 'independence', became *more* of a colony of the capitalist centres in the West than she had ever been in the previous era when she was a colony of feudal Spain and Portugal.

This process of underdevelopment by integration into the Western capitalist economy generally occurred in the following ways.

(*a*) Through the establishment of a *tax system* ostensibly designed to make the colony 'self-financing' – this had far-reaching social and economic consequences in that it introduced vast areas of the world to the money economy, and so subjected these areas to the dictates and the market fluctuations of the capitalist centres in Europe. In order to earn the white man's taxes, villagers either had to offer their labour for wages in the white man's settlements, that is on their plantations and mines, or they had to grow cash crops for the export market. In both cases the effects were similar in that the self-sufficiency of the village as a socio-economic unit was broken. Examples of such infamous tax systems operative in different parts of the world are the 'hut tax' in South Africa, the 'poll tax' or 'head tax' in other parts of Africa, the 'mita system' in Latin America, the 'cultuur' system in the Dutch East Indies, and the 'permanent settlement' schemes in India.

(*b*) Through the direct organisation of *the production and the marketing* of the colonies' raw materials and agricultural produce necessary for the further industrial expansion of the West – this indeed is the essence of colonialism, or as Lenin called it, the imperialist stage of capitalism, namely that the centre organises and monopolises the production of the export produce of the periphery.[32] The colonial period saw the granting of concessions for mining and the lease of land for plantations to subsidiaries of giant metropolitan industries. This ensured the vertical integration of the major capitalist enterprises, which consolidated their monopolistic power positions, and this, in turn, facilitated political de-colonisation without economic de-colonisation at a later date.

(*c*) Through the *organisation of market outlets* for the finished products from the mother country – the colonial domestic market was quite simply protected against competition from third countries. It is in this period that the massive expansion of industry in the West lowered the prices of manufactures to a level where they presented the 'kiss of death' to the already withering local crafts. Furthermore, colonial policy on occasion would issue commercial decrees designed to eradicate indigenous industries, or indeed the desire for them altogether.

(*d*) Finally, through the *monetary* adjustment of the colony to the mother economy by means of the simple but brilliant expedient of the creation of currency zones, for example 'the sterling area', 'the franc zone', the 'escudo zone', and so on. The colonies were directed to use the currency of the mother country as a reserve currency for a part or all of their monetary reserves, to make their foreign payments in that reserve currency, and to keep their own currencies stable in relation to the mother currency rather than any other world currency. As Magdoff points out,[33] the control potentials implicit in the use of a particular currency as a reserve are enormous.

(1) It facilitates control over the trading arrangements of the colonies. And up to this day, because of these currency zones which have continued after de-colonisation, the trading patterns of under-

developed countries are not only characterised by a concentration of export commodities but also by a concentration of trading partners. In other words, the mother country continues to be the most important trading partner.

(2) The enforced use of the mother currency as reserve currency allowed the mother country to finance a trade deficit which she may have had with her satellites (as was especially the case with Britain during the Second World War).

The subjection of the colonies to an imposed international division of labour encouraged 'monoculturisation' which generated *poverty*, and which sowed the seeds of economic and political instability for many years to come.

Whether under the pressure of the colonial system of taxation and administration, or more directly from the white settler's greed, as well as resulting from the incorporation of the colony into the world capitalist system, ever larger areas of arable land were brought under the cultivation of one or two particularly suitable crops, be it sugar, coffee, cotton, palm-oil, tea, sisal or bananas. With virtually all arable land cultivated for cash crops, precious little was left for the native population to grow their staple food. In good years, the money earned with the cash crops would cover the importation of staple food; in bad years it quite simply would not. And, with agricultural commodities like sugar, cocoa, coffee, cotton, rubber and bananas being sensitive to fluctuation in world demand, a kind of 'boom and bust' economy grew up, which further eroded the livelihood of the people. In boom periods more land would be brought under cultivation of the cash crop concerned, only to be abandoned in the next bust period, which – needless to say – was itself the very outcome of the increased cultivation prompted by the boom. Because of the relative inadaptability of agriculture, such *extensive* forms of agricultural production generally made for an extremely inefficient use of the soil so that still less land could be used for staple food production. De Castro's *The Geography of Hunger* gives an excellent account of how poverty accompanies this one-track exploitation of the soil.[34] While huge fortunes were made on the sugar plantations of North Brazil and pre-Castro Cuba, the native population was reduced to absolute starvation.

The lop-sided economic structure based upon a narrow primary-production structure was the single most important and enduring colonial heritage. In 1970, about ten years after the last move for formal independence a U.N. report observed that at that time 'almost 90% of the export earnings of the developing countries derive from primary products. Moreover, nearly half of these countries earn more than 50% of their export receipts from a single primary commodity. As many as three-quarters of them earn more than 60% from three primary products.'[35] Thus, in one sense one may argue that the result of colonialism was a further underdevelopment of the conquered territories. Yet, viewed from another perspective, colonialism yielded quite a contrary result: the organisation of primary production and the guaranteeing of markets for the finished products from the mother country implied a transfer of capital, technology and Western modes of socio-economic organisation to the colonies. Marx himself, perhaps more than any of his followers, acknowledged this when he observed that 'England has to fulfill a double mission in India – one destructive, the other *regenerating* – the annihilation of old Asiatic society, and the laying of the material foundations of Western society in Asia' (emphasis added).[36] In other words, westernisation and modernisation had begun, and this process of modernisation brought in its wake yet another series of problems 'typical' of contemporary developing societies. We shall look at these in the following chapters of this part of the book.

NEO-COLONIALISM

Lastly, the third phase in the dialectic of development and underdevelopment can also be seen as a logical outcome of the preceding phase: once the monopolistic control over the production and the marketing of the periphery's export produce had been consolidated, and the market outlets had been guaranteed, there was no longer a need for direct political control by the mother countries over their colonies. Political de-colonisation proved a budget-saving as well as a humanitarian act. International law, a useful European invention, recognises individual and corporate property rights across national boundaries, so that the big industrial corporations and the major commercial banks could continue to regulate the ex-colonial economies without the military or the police support of their own respective national governments. The term 'neo-colonialism' refers to this retention and the further extension of economic control and influence by the ex-colonial powers after they had surrendered political state power.

The export of capital from the 'centre' to the 'periphery', which was, as we have seen above, such a dominant feature of the colonial period, and one which helped to tie the colonial economy to that of the mother country, continues today. Ironically, one reason for this is that precisely with the assertion of independence came the desire for 'development', the model and the means for which the de-colonised peoples continue to look for in the West. So the West's market outlets remain guaranteed. Westernisation during the colonial period has helped to gear the colonial peoples' tastes and their material needs to a level of satisfaction of those needs well beyond the productive capacity of the new nation. The reliance on Western 'patent solutions to basic human needs'[37] ranging from trucks and cars through Coca-Cola and canned fruits to hospitals and schoolrooms, persistently thwarts indigenous development efforts, as the demands for these western artefacts outstrip the new nations' export earnings.

One way of understanding how this third phase in the history of development and underdevelopment should make for further *under*-development of the neo-colonial territories, is given in Raoul Prebisch's analysis of the deterioration of the terms of trade for the poor countries.[38] Recognising today's international division of labour, Prebisch sought to disclose the causes of the price increases of western manufactures, and the simultaneous price *de*creases of the raw materials and the foodstuffs which are the export produce of the ex-colonial territories. Briefly, his argument was this: after industrialisation had started in Britain, other European countries followed suit. This made for tough competition, and in order to reduce the cost of production wages were kept low at first. A little later, however, in the nineteenth century, trade unionism and socialism brought about the emancipation of the working classes. From then on, to meet increasing wage demands, the prices of manufactured goods showed an ever-greater tendency to rise, whilst at the same time a deliberate attempt was made to keep the prices of raw-material imports as low as possible, as an alternative way to reduce the cost of production. This the capitalists of the West were in a position to do, since they themselves, as we have seen earlier on, either owned or controlled the production and hence the prices of primary commodities in the colonial, and later ex-colonial territories. A phenomenon related to the process of development in the West has been the tremendous progress in medical aid. The fruits of such advancements moved to the colonial territories, where they upset the balance of the population, thus creating a permanent over-supply on the labour market. This considerably facilitated the pursuit of low-wage policies. In short, Prebisch's argument is that, whereas increases in productivity in the centre were taken up by higher wages, so that the prices of manufactured goods increased, exactly the opposite took place in the periphery – there the increase in productivity led to a decline in the prices of primary commodities.

Besides these *politico-economic* reasons for the deterioration in the terms of trade, Prebisch also cited three *technical economic* reasons for the further deterioration of the terms of trade. First, as it happens, primary products are subject to substitution by synthetics; secondly, agricultural commodities are typically demand-inelastic (Engels's Law), and thirdly, technological progress makes for an increasingly unfavourable ratio of raw-material inputs in manufacturing production.

In the 1960s, Prebisch's thesis became by far the most popular version of the development of underdevelopment theory. His became, in fact, the founding theme of the U.N. Conferences on Trade and Development, both in 1964 and in 1968. Many well-intended and progressive aid and Third World movements in Western Europe and North America drew their inspiration from the 'unequal trade' argument,[39] and many – not always successful – attempts were made to cajole housewives into buying cane sugar instead of beet sugar, into drinking 'real' coffee instead of Nescafe, into boycotting various brands of tea, and generally into being sensitive toward products from Third World countries. At a more sophisticated and international level, the U.N. Second Development Decade was launched with ambitious plans for price protection of primary products, and many schemes ranging from commodity agreements and buffer-stock arrangements through to international compensatory finance were proposed, and some implemented.[40] Last but not least, developed countries were urged, and with moderate success, to open their domestic markets to manufactures from the underdeveloped ones, a proposal that also stemmed directly from the Prebisch analysis.

Of course some economists had criticised the Prebisch analysis, and especially his data, and they had argued that there never really had been a *sustained* deterioration of the terms of trade for primary producers this century, and that data such as those produced in the Prebisch report would never have surfaced had different years been selected as base years (see Table 4.1, p. 89).[41] Other economists quickly countered this by pointing out that the data on primary-product prices are aggregate figures, which lump together earnings of developed and underdeveloped primary producers alike, and that whilst the developed commodity exporters such as the United States, Australia and Canada had managed to improve their terms of trade, the underdeveloped ones, politically weaker, had suffered further deterioration.[42]

And so the debate wore on, but by and large the 'unequal trade' argument gained ground in all but the most conservative of public opinions and political organisations, until the shattering experience of the commodity boom in 1973. For many well-meaning aid and Third World movements, the near doubling of almost all commodity prices had a profoundly disorientating effect on their campaigns, and when the oil-producing countries through a very effective political/economic cartel (OPEC) quadrupled *their* prices, and – albeit for political reasons – added insult to injury by initially refusing to recycle their enormous fortunes back into the coffers of western capitalism, even convinced socialists and faithfuls of the Marxist articles on imperalism were thrown into confusion.

These recent events, as well as the earlier criticisms of the Prebisch analysis, demonstrate the vulnerability of an approach which makes too much out of the symptoms of a disease, but which does not adequately analyse the nature and the causes of the disease.

Underdevelopment in the contemporary world is a particular form of capitalist development, namely *dependent capitalism.* Both the famines in India and Bangladesh, and the fabulous fortunes of the Arab emirates are symptoms of the same dependency. Deteriorating as well as rising commodity prices issue from the same syndrome. In the course of their cyclical movements the capitalist centres of the world transmit pulsations of contraction and expansion to their peripheries. Thus the recent commodity boom is largely explicable in terms of a preceding industrial expansion in Europe, Japan and North America which relayed a rising demand for primary commodities to the peripheries. But every boom is followed by a bust, and the world recession which we are now experiencing will no doubt be followed by a glut in commodity prices. And, unless these primary-exporting countries which benefited temporarily from the boom with a real net increase in foreign exchange, have succeeded in transforming this increased wealth into an expansion of the productive capacity of their own economies, they cannot expect any lasting effects from the boom. And it is exactly the *dependent* type of production structure of the great majority of the underdeveloped countries which prevents them from materialising the gains of their increased export earnings. For, first, as we observed earlier, theirs is a lop-sided production structure, producing one or two or three commodities only (see Table 4.2, pp. 90–91), so that many underdeveloped countries are themselves significant importers of primary commodities. Much of their increased export earnings during the boom has been offset by the increased costs of their imports. It is conservatively estimated that imports of agricultural commodities and fertilisers by non-oil-producing developing countries rose by at least five billion dollars in 1973 over 1972 imports of six billion dollars. Thus, well over one-half of the gains of all developing countries (excluding oil producers) estimated at about eight billion dollars were lost in increased import costs.[43] Secondly, the initial disparity in standards of living between industrialised and non-industrialised nations will – in the long run – continue to disequalise the terms of their trade, since the products of the latter always form but a cost component of the products of the former. In other words, in so far as the boom in food and raw-material prices raises the prices of manufactured goods, as it inevitably does through higher wages and costs of materials, a further reduction of the benefits from high commodity prices must be expected. Thirdly, the foreign-exchange reserves are necessarily held in dollars or sterling, another reflection of the imperialist concentration of the world capitalist system. Under the present inflationary conditions of the major capitalist countries any temporal lag between export earnings and import costs implies a real devaluation of the foreign assets of underdeveloped countries. Finally, quite a proportion of underdeveloped countries' export commodities is produced by multinational corporations or other foreign investors. This is especially true of the extraction of minerals. As Singer has pointed out, for this portion of the commodity boom much of the benefit will show in terms of increased profits for foreign investors rather than for the exporting countries concerned.[44]

In short, a temporary relief of certain symptoms of the disease of underdevelopment should not blind us to the continued existence of the disease. With the exception of a handful of tiny Arab oil-producing countries who, thanks to an effective political alignment in OPEC, can now, as Emmanuel puts it, 'dispose of a volume of royalties so disproportionate to their population as to enable them to live without working, and consume without producing',[45] the rest of the Third World is still very much at the mercy of the dictates and the fate of the capitalist centres of the world, and because of their external dominance and dependency they not only continue to resemble one another in important economic, social and political respects but they also continue to be underdeveloped.

Such at any rate is the position of the neo-Marxist theorists of imperialism who see the continued dependence of Third World countries upon the world capitalist system as a recipe for *further underdevelopment*, and who consequently advocate a radical break with that capitalist system as a necessary condition for economic independence, and, hence, development.

What is problematic, however, is not just the commodity boom, but the more fundamental empirical question of whether or not 'economic development' has taken place since formal independence, say roughly in the last 20 years, in those countries of the Third World that have continued to live in symbiosis with world capitalism. 'Bourgeois' economists have tended to view the experience of capitalist Third World countries, certainly since the early 1960s, rather more optimistically than their Left-Wing colleagues. They base their optimism on the economic growth rates of 'selected' developing countries, and more particularly on the impressive rates of expansion of the manufacturing sectors of these same selected countries, the latter, of course, being taken as a sign of rapid industrialisation. And, since these same selected countries also appear on the list of those who have had a most liberal attitude *vis-à-vis* the importation of western capital and technology, their argument is that close ties with the capitalist world in fact *promotes* economic development.[46]

Normally one does not, of course, expect Marxist and 'bourgeois' economists to see eye to eye with each other over such fundamental issues as the organisation and the consequent direction of economic life. So one does not as a rule bother too much when one finds their economic predictions wildly at odds with each other. But when – as is the case at the moment – it is not just their predictions but their interpretations of the experience of the past twenty years which are diverging, and when, moreover, these interpretations take the same statistical data, collected by the same reputable and reliable international organisations such as the United Nations and the World Bank, as the factual basis for their contrasting views, then a closer examination of their arguments is called for.

This debate, which was intelligently pursued in several issues of the *New Left Review* under the appropriate titles of, respectively, 'Current Myths of Underdevelopment' and 'Current Myths of Development',[47] is all the more interesting to us because it brings into focus the very controversy which is the underlying theme of this book, that is 'diffusion' versus 'domination'.

The *diffusionist* view put forward by Bill Warren contends that, contrary to widely held beliefs, marked economic development by way of *industrialisation* has been achieved in the Third World taken as a whole; that this industrialisation has occurred as a result of the transfer of capital and technology from the West; that this transfer of capital and technology has encouraged the spread of capitalist social relations and productive forces (that is commodity production in agriculture, the purchasing of labour power and raw materials by capital in industry, and the resulting continuing reproduction of capital); that this spread of capitalist social relations and of capitalist productive forces has in turn encouraged indigenous, autonomous and national economic interests to assert themselves; that formal political independence complemented by a shrewd 'playing-off' of imperialist rivalries by Third World countries has added further impetus and bargaining strength to these emerging 'nationalist capitalisms'; and that, as nationalist capitalisms grow, imperialism (that is the domination and control by the capitalist centres over their peripheries) gradually withers away.

Here in a nutshell we have the economist's contribution to the diffusionist approach to modernisation which we discussed at such great length in the first part of the book: capitalism is seen to spread out from the original centre into the peripheries, where it reproduces the economic, social and political conditions under which it needs to operate. In this process of reproduction the control by the centre over the peripheries, which is a necessary condition of the early phases of diffusion, gradually tapers off as the reproduction of capitalism becomes more complete. Imperialism declines as capitalism grows, much in the way humans reproduce themselves, nurse, attend and direct their children until these, too, are grown up and become independent of their parents. Warren does present empirical 'evidence' of sorts for each of his connected theses.

First, and most importantly, the assertion of *rapid industrialisation* in the Third World as a whole, is backed up by no less than five statistical displays showing:

(*a*) that Third World manufacturing output, at 7 per cent per annum between 1960 and 1968, grew 1 per cent faster than that of the advanced capitalist world at 6 per cent.

(*b*) that the percentage share of Third World manufacturing in total world manufacturing increased, reducing the 1937 ratio of nine to one (advanced capitalist world share versus Third World share) to seven to one in 1959;

(*c*) that the manufacturing growth rates of some twenty-two 'selected' countries were sustained over a period 'longer than any one previously recorded';

(*d*) that the proportion of gross domestic product accounted for by manufacturing rose from 14·5 per cent in the period 1950–4 to 17·9 per cent in the period 1960–4, while in the developed capitalist countries during 1960–4 manufacturing contributed 31·3 per cent of G.D.P. Hence, the proportion in underdeveloped countries is already one-half that of the developed ones; and

(*e*) that, although the proportion of active population engaged in manufacturing is not commensurate with the proportion of manufacturing output to G.D.P., 'progress has been made in some important Third World countries, which, thanks to very substantial industrialisation, have characteristics far different from those typical of underdevelopment'.[48]

Secondly, to support the theses about the emerging strength of nationalist capitalisms and the corresponding weakening of imperialist ties, Warren calls our attention to a whole range of current sophisticated policies and practices of Third World countries all of which indicate an improvement in their bargaining position *vis-à-vis* foreign *extractive industries:* such as joint ventures, local equity participation, service contracts and outright nationalisation. According to Warren, the sophistication of these policies lies in the fact that they combine increasing indigenous control with continuing access to the most advanced methods and technology abroad. He admits that with respect to local subsidiaries of foreign *manufacturing* firms (which, however, are less important in size and are of relatively recent origin), the host governments are still at a stage where they want to attract more foreign investment and therefore are 'nearer the beginning of the cycle of negotiation, renegotiations and increasing control, sometimes ending in outright nationalisation, through which the resource industries have passed'.[49] But here, too, Warren remains optimistic about the trends toward greater control and the moulding of operations to suit national development aims in view of the tough competition between the foreign giant industries, *vide* the increasing share of West German, Japanese and U.S. direct and portfolio investment in the Third World as against the declining share of the old imperialist powers.*

Warren's position – which to be fair seems representative of that of a respectable and authoritative club of development experts – has been countered by Marxist economists in two main contexts. The first deals with the *empirical facts* of the economic development and industrialisation allegedly taking place in the Third World. The second deals with the interpretation of the apparently growing independence and autonomy of the national economies of those Third World countries that seem to be developing, and more especially of the twenty-two selected countries which Warren, following Chenery,[50] adviser to the President of the World Bank, has chosen as examples.

As regards the economic facts of development, the whole of Warren's – and generally of bourgeois economists' – 'evidence' about development in the Third World hinges on the presentation of *absolute* growth rates both of gross domestic product and of manufacturing output. These growth rates are indeed impressive, but they dwindle into insignificance when *calculated per head of population.* And surely 'economic development', claim Marxist economists, is only meaningful when it refers to a rise in the productivity and in the standards of living of *people* rather than of *nations.*[51] When we calculate the growth of output per head of population, the economic performance of developing countries does *not* compare favourably with that of the advanced capitalist nations. Warren dismisses this as applying a 'too demanding criterion of performance',[52] and he squarely puts the blame on the unprecedented population growth rates of the underdeveloped countries. But Marxist economists rather attribute this poor performance to the nature of foreign-induced, highly capital-intensive industrialisation in the Third World which, whilst allowing a dynamic growth of the industrial sector, has not permitted the labour-absorption rate of this sector to keep pace. In fact, in at least one of the selected countries, industrial growth is reportedly accompanied by a net decline of industrial employment, whilst in others the growth of employment of labour merely lags behind (see Table 4.3, p. 91).[53] What this means is that the optimistic rates of growth of both total output, and of the manufacturing sector in developing countries since their formal independence, merely reflects the fact that 'nations' are developing,

* Warren reports that whereas U.S. direct and portfolio investment (including re-investment) in the Third World as a whole grew at an annual average rate of 15.4 per cent between 1964 and 1968, Japan's investment during that same period grew at an annual rate of 32 per cent and Germany's at a rate of 50 per cent.

but that the people in them are not. More and more the developing economies show a structural distortion, providing highly rewarding employment for a small minority, and a very large amount of barely productive poverty-line employment, or no employment, for the masses. If economic development is taking place at all, it is associated with increasing inequality. A systematic study of the impact of economic growth on income distribution in developing countries leads Adelman and Morris to conclude that 'development is accompanied by an absolute as well as a relative decline in the average income of the very poor. Indeed, an initial spurt in dualistic growth may cause such a decline for as much as 60% of the population.'[54]

Another important indicator of economic development, be it of 'nations' as opposed to 'peoples', in Warren's analysis is the increasing *share* of manufacturing activity in total domestic production. Such an indicator, however, is misleading for it may reflect, as the Marxist economists claim it does, an increasingly *low* productive performance of the non-manufacturing (that is agricultural) sector in the Third World.[55] If the increase in the proportional share of the manufacturing sector in the G.D.P. of the Third World countries compares favourably to that in the advanced capitalist countries, it is rather because Warren omits to tell us that thanks to the *industrialisation of agriculture* (that is mechanisation) in the *advanced countries*, productivity of agriculture *there* has leapt still further and further beyond the productivity of agriculture in the underdeveloped countries. As Emmanuel reports:

> Whilst in the period between 1959–1970 the productivity of labour of manufactures in the Third World has increased from index 100 to index 129 (at constant prices) the productivity of labour in agriculture in the O.E.C.D. countries leapt from 100 to 185, and in the U.S.A. to 171. As for the productivity of labour in manufactures in the advanced countries this has increased in the same period from 100–160.
>
> Clearly, therefore, it is not by transferring its factors from agriculture to industry that a country develops, but by mechanising and modernising *both* of these sectors. The superiority of the O.E.C.D. countries over the Third World does not consist in the larger share occupied by manufactures in their national production, but in the fact that both their manufactures *and* their agriculture are on a far higher level than those of the Third World.[56]

In other words, relative to the developed countries the underdeveloped countries continue to stagnate.

Finally, the impressive expansion of *exports* of manufactures from the Third World, which is a last trump-card in Warren's list of development indicators (after all the proof of the pudding is in the eating, in this case by the already industrialised nations) turns into yet another 'myth' when we learn that over one-half of these exports come from five developing countries only, of which Hong Kong accounts for 21·5 per cent (in 1968).[57]

This last observation leads us directly into a critical examination of the relative position of the twenty-two selected countries, upon whose exemplary performance Warren – following Chenery – bases his entire argument. Generally speaking, it seems to me, the statistical practice of 'averaging' the development performance of all Third World countries into figures representing the performance of the Third World as a whole is a most misleading exercise and one that should be abandoned forthwith. In 1970, the total population of the eighty or so developing Third World 'market' economies usually included in these 'averaging' exercises totalled about two billion of the world's 3·5 billion people. As Table 4.4 (p. 92) shows, of the twenty-two countries which Warren has selected as examples, nine were *very* small, with a combined total population of no more than 20 million (a very small number indeed); another ten were middle-sized with a combined total of 213 million, and only three had really large populations of 50, 80 and 130 million respectively. Their manufacturing growth rates vacillate around 10 per cent. Averaging the industrialisation performance of *all* Third World market economies, and arriving at a proud figure of 7 per cent growth per annum, which is what Warren does to 'prove' his industrialisation theory, makes one realise just how *poor* the performance of the rest of the Third World market economies, those representing three-quarters of the peoples involved, must have been in the period under discussion. In short, bourgeois economic presentations of 'facts' of development in Third World countries do not just hide the internal reduction in the standards of living of the masses of the people *within nations*, they also hide the wide disparity *between* Third World nations, whereby the stagnation of the nations with large populations is often obscured by the seemingly impressive performance of a handful of nations with small populations.

With respect to the second issue, namely, the interpretation of the Third World's allegedly increasing economic 'autonomy' and 'independence' the Marxist critique is less clear cut. Unfortunately, Marxist economists are themselves divided over the important question of the nature of contemporary imperialism. As a consequence, they disagree amongst each other about the *manner* in which imperialism is to be held responsible for the continued underdevelopment of the Third World. Whilst one view argues that continued underdevelopment and stagnation is the result of the *concentration* of world capitalism within the already advanced world, the other view rather emphasises the *character* of its current *expansion* as a cause of further underdevelopment. Although the two views are not necessarily incompatible, the difference in their perspectives does make it difficult to amalgamate them. I shall therefore briefly present them one after another. The former view, expressed by Emmanuel, contends that poor countries remain poor, because being poor they are unattractive to foreign investment.

> Apart from raw materials and certain agricultural products which have to be sought where they can be found, the movement of capital is not an increasing but a decreasing function of differences in incomes. . . . The advanced countries are nowadays too rich not to be able to absorb themselves, without difficulty, all the new capital

that is formed in them, and the underdeveloped countries are too poor to offer attractive investment prospects to this same capital, apart from their few import substitution industries. They are even so poor, that they dispatch to Switzerland part of their own national surplus. All this, in turn, keeps them poor, or makes them even poorer.[58]

Thus, in this view, the majority of the underdeveloped countries are underdeveloped not because they have been invaded by foreign capital but because they have been starved of this capital. As this is the case with the greater number of the underdeveloped countries, Emmanuel further contends that only the few relatively small underdeveloped countries that have opened up most liberally to foreign investments have a chance of succeeding within the system, as a kind of marginal client state, so to speak. But capitalist development remains an impossibility for the larger part of the underdeveloped world, precisely because of the concentration of capital within the already advanced world.

Emmanuel's position ties in with the widespread belief that there is currently a decline in overseas investment in raw-material production, a switch of investments into manufacturing subsidiaries of giant corporations, and consequently a redirection of capital flows away from the underdeveloped lands and into the already developed ones. Naturally this contention, too, has its statistical backing. Barrat Brown, who observes the same trend in his recent work on the *Economics of Imperialism*, produces a table which he says (see Table 4.5, p. 93)

shows that the direction of U.S. investment is not now so much to excolonial or underdeveloped lands as to other industrially developed states. This is true for all the main exporting countries. About half of all their capital exports went to each other in the 1960s. At the same time a high proportion of investment income still arises from underdeveloped lands and particularly from oil investments. There is once again a switch of investment taking place in the establishment by manufacturing firms of subsidiaries in the more and not the less, industrially developed countries.[59]

Curiously, Raymond Vernon in *Sovereignty at Bay* presents equally up-to-date tables with data on foreign direct investment of U.S. enterprises in manufacturing subsidiaries which show, as he says, 'unmistakably' exactly the opposite trend. I quote: 'From their pre-war and wartime base U.S. business commitments grew rapidly in all areas of the world but the rate of growth, on the whole, was especially rapid in the "new" areas of the world where strong forces were at work pushing U.S. enterprises outward into unfamiliar territory.'[60]

As a matter of fact, if we recalculate Vernon's figures into the respective percentage share of the advanced countries' and the underdeveloped world's intake of U.S. foreign direct investment, we see that their respective shares of U.S. capital has remained very nearly stable throughout the period 1929–69 (see Tables 4.6 and 4.7, p. 94). Whichever the correct interpretation of these statistical data, it is clear that – as sociologists at any rate – we shall be none the wiser from these economic statistical quibbles.

The *critical* point, rather, and one which is also brought forward by the second Marxist position against Warren's view, is that during the neo-colonial phase of underdevelopment, the degree of control and domination of the Third World is no longer an immediate function of the size of exported capital from the capitalist centres to the peripheries. The secret behind this riddle of modern imperialism lies in the *organisation of giant transnational* corporations, and *their practice to associate themselves in joint ventures with local – private as well as state – capital in underdeveloped countries.* Magdoff has estimated that during the period 1957–65, of the 84 billion U.S. dollars used to finance the expansion and the operations of U.S. direct foreign investments,

> only little more than 15 per cent came from the United States. The remaining 85 per cent was raised outside the United States, 20 per cent from locally raised funds and 65 per cent from the cash generated by the foreign enterprise operations themselves . . . the pattern is similar for rich countries and poor countries. If anything, the U.S. capital contribution is less in the poor countries than in the rich ones.[61]

In contrast to the colonial phase of underdevelopment, it is no longer the imperialist rivalries between the capitalist *states*, but that between the modern *transnational corporations* which determine the economic, the political and indeed the social reality of the world today. The competitive struggle between these giant firms (some 200 of whom control between them over one-half of the total world's output); their need to capture raw-material sources, and their drive to monopolise existing and potential markets, is what decides the allocation of resources, the patterns of production, the techniques of production, and the relations of production *on a world-wide scale*. The most conspicuous feature of the organisation of the multinational enterprise is the fragmentation of the process of capitalist production such that 'each industrialising colony or semi-colony partakes of a part of the industrial process but not of the whole'. Therefore,

> Much of what Warren has euphemistically referred to as 'industrialisation' has been in large part the development of 'assembly plant' operations. Hence, to assume equivalence of capitalist industrialisation within imperial centres and the Third World is to overlook essential differences in the structure of industry and levels of development of productive forces, as well as the significantly different class structures that mediate international economic relationships.[62]

The increasing alignment of the national bourgeoisie and the governments of the underdeveloped countries on the one hand, with the new centres of international capitalism, for example the multinational corporation, on the other, is the key to the understanding both of the structure of the economy and society in Third World countries and also of their further stagnation and underdevelopment. The very same practice which the diffusionist Warren sees as a sign of sophistication and independence (namely joint ventures and the participation by local capital in foreign enterprises) on the part of the national bourgeoisie of underdeveloped countries, the neo-Marxists critically evaluate as yet further *dependence* in a new cloak.

In the next chapter we shall have a look at the class structure in underdeveloped countries which mediates these 'international economic relationships'.

As this is hardly a textbook on the economics of imperialism, nor on the multinational enterprise, I shall conclude this chapter on the development of underdevelopment by merely briefly mentioning a *range* of current practices by multinational firms which both enhance their domination over the economies they invade, and which simultaneously reduce these economies' potential for 'independent' development, that is which further underdevelop them.

The practice of transfer pricing, or double accounting

It is often argued that a major advantage of foreign investments in underdeveloped countries is that it presents an immediate inflow of foreign currency, which can ease the balance-of-payments constraints. And, provided national governments are alert and 'sophisticated' to tax profits and tighten up exchange regulations to limit the repatriation of profits, and to ensure the ploughing back of these profits into the coffers of the host economy, then surely all is well. But this is exactly what many underdeveloped countries governments *have* done, indeed it is an act of independence for which they have been loudly praised. And yet, when we look at the inflows and return flows of direct investment of capital between the United States and the underdeveloped world, we notice that between 1950 and 1965 almost three times as much money was taken out as was put in (see Table 4.8, p. 95). The reason for these large return flows despite the underdeveloped countries' 'sophistication' in their negotiations with the large foreign firms, is that in today's modern imperialism most of the profits are *repatriated not as profits but by means of what is called transfer pricing or double accounting.* This is the practice of overpricing by the parent company of the home-office sales to their subsidiaries and of underpricing its purchases from them, so that a larger part of the taxable profits are hidden under cost items.[63] Vaitsos, in his contribution to Bernstein's book on *Underdevelopment and Development*, makes the following pertinent calculation: 'Defining as effective returns to the parent company the sum of reported profits of the subsidiary, royalty payments and intermediate product overpricing, reported profits of a sample of the Columbian pharmaceutical industry constituted 3·4% of effective returns, royalties 14·0 per cent and 'overpricing' 82·6%.'[64]

The protection by patents

The overpricing of intermediate products, or of the parent company's technology, occurs, or is made possible, because of the monopolistic or oligopolistic nature of the market for intermediate products and technology, protected as they are by an international 'con-trick' called 'patents'. For example, a licensee of Datsun Motors is bound by patent to import components from the licensor since the technology embodied in the chassis and in the assembling parts of Datsun cars requires specific Datsun components. Tugendhat reports that when Mexico some years ago tried to cut imports and encourage the development of local industries by forcing the U.S. car companies manufacturing in that country to obtain a higher proportion of their components in Mexico, and to this end forbade certain imports, the companies retaliated by adding the mark-up they would have got on the forbidden imports to those components which they were still allowed to bring in.[65] Although figures on these issues are very hard to come by, it seems reasonable to suggest that already a high proportion of trade is accounted for by transactions between affiliates of multinational corporations.*

What this suggests is that, increasingly, 'profits' will be made not from final products in consumer markets, but from cost-pricing of the technologies involved in producing these goods. Since these technologies 'belong' to the parent company, which is invariably located in the advanced world, and since in the 'multinational synergy' the ultimate profit maximisation of the parent company is the only ultimate goal of the application of 'economic rationality', it can be reasonably deduced that the subsidiaries of these companies in the underdeveloped countries, and the host underdeveloped countries themselves, will ultimately be the losers. Thus the new economic ties between rich and poor consist of patents, technologies and licences.

The calculation of equity capital

Modern international capitalism, or neo-imperialism, is – as suggested before – marked by a high degree of control over dependent economies coupled with a comparatively low amount of actual cash investment. What happens is that most corporations calculate know-how, blueprints, patents, expertise of management, and so on as perhaps as much as one-third of the value of their total investment, and then supply another third only in equity by providing machinery and equipment. Hence, a good share of the assets owned by, for example, U.S. firms overseas does not represent cash investment, but a valuation by the company of their knowledge, trade-marks, and so on.[67] Such practice, of course, presents ample scope for fiddling with local tax legislation in order to help the repatriation of profits. Furthermore, in cases of nationalisation and state participation there is a much higher compensation to be gained from such valuations.

Control over market and market conditions

The power of the top management of the parent company of any international concern lies in that it reserves the right to decide how

---

* This in view of the fact that about one-quarter of both U.S. and U.K. exports in 1964 were sales from multinational companies to their affiliates abroad.[66]

and where its products should be manufactured, where they should be sold and at what price. These four rights give the parent company all the power that is needed to control markets and market conditions. If a multinational group wants to hit a rival hard in a particular market it may subsidise its own local subsidiary in order to enable it to launch and sustain a price war. In other words, the subsidiary may be ordered to act against all the canons of economic rationality for a long time. Hardly a way to make it contribute to the national economy of the host country![68]

Neighbouring countries with different foreign company laws and tax legislation prove an easy and helpless prey to the multinational. If one has a higher tax on profits, the parent company may simply increase prices, whilst directing its other subsidiary next door to invade the market with lower prices.

Defensive investments

One might sum up the characteristic face of neo-imperialism by saying that it is after power first, profits later. The desire for monopolistic power (which of course ensures profits in the long run) is a more overriding motive for its pattern of foreign investments than any immediate returns on these investments.[69] Multinationals may feel they 'have to' invest or set up a subsidiary to protect their share of the market in a corner of the world even though this may be a very costly thing to them initially. Often trade barriers set up by governments of underdeveloped countries to protect their own infant industries will motivate a multinational to invest in that country to protect its interests in the market, that is on the other side of the trade barrier.

Patent suppression

Again, often a multinational may steer its way into an underdeveloped country under very attractive conditions to the local government, in return for permission to take out a series of patents for the production and sale of certain products, only to develop one or two of these and suppress the rest, that is the other patents, because its subsidiary in a neighbouring country may operate under more favourable conditions (namely in terms of tax legislation) and so be given the opportunity to move in.[70]

Control of exports

Multinational parents not only decide when and how and where their products should be manufactured but also where they should be sold. Its subsidiaries are often not free to choose their own export markets, or to compete for them with each other. Very often, even in cases of state-owned, state-participated, or so-called joint ventures, the contracts with the parent company explicitly prohibit exports of products manufactured with the use of imported technology.

Organising the over-all production process

In the multinational synergy, the parent company is careful to fragment the production process so as to hand out the various aspects of the production process in accordance with local factor endowments. Thus foreign investment in underdeveloped countries tends to concentrate (besides in the natural choice of extractive industries) in light consumer industries, due to the availability of cheap labour as well as the low level of skills required, and reserve its producer-goods industries for the more advanced countries. Expansion of these light consumer industries invariably leads to a heavy balance-of-payments pressure because of the increasing need for foreign capital goods.

Table 4.1

Quantity and Value Export Indices, and Terms of Trade[1] *1938, 1948, 1950–71*

Year	1950 = 100						1963 = 100					
	Developing countries			*Developed countries*			*Developing countries*			*Developed countries*		
	Quantity	*Value*	*Terms of trade*	*Quantity*	*Value*	*Terms of trade*	*Quantity*	*Value*	*Terms of trade*	*Quantity*	*Value*	*Terms of trade*
1938	94	30	71	84	41	109	56	19	80	36	15	100
1950	100	100	100	100	100	100	59	62	112	43	36	92
1954	108	114	100	131	149	100	64	70	112	56	53	92
1958	127	128	93	163	192	104	75	79	104	70	68	95
1962	159	149	87	216	257	109	95	93	98	93	92	100
1966	200	199	90	302	378	109	118	122	101	130	136	100
1969	250	252	90	402	519	110	147	156	101	173	187	101
1971[2]	283	306	90	467	669	110	167	190	101	210	241	101

NOTE: Includes petroleum.

1. Unit value index of exports divided by unit value index of imports. 2. 1971 figures are provisional.

SOURCE: United Nations, *Trends in Developing Societies* (World Bank, 1973).

Table 4.2

Leading Export Commodities of Underdeveloped Nations (based on 1967 trade data)[1]

Country	*Number of Leading Export Commodities*	*Export of leading commodities As per cent of total exports*	*Leading export commodities*
Argentina	4	61	Meat, Wheat, Corn, Wool
Bolivia	1	63	Tin
Brazil	4	58	Coffee, Iron Ore, Cotton, Cocoa
Cameroon*	3	65	Cocoa, Coffee, Aluminium
Central African Republic	3	90	Diamonds, Coffee, Cotton
Ceylon	3	89	Tea, Rubber, Coconut
Chile	3	85	Copper, Iron Ore, Nitrates
Columbia	2	69	Coffee, Oil
Congo, Democ. Republic*	4	74	Copper, Tin, Diamonds, Coffee
Congo (Brazzaville)	2	76	Wood, Diamonds
Costa Rica*	2	60	Coffee, Bananas
Dominican Republic	5	91	Sugar, Coffee, Cocoa, Bauxite, Tobacco
Ecuador*	3	84	Bananas, Coffee, Cocoa
Ethiopia	4	84	Coffee, Hides and Skins, Cereals, Oil seeds
Gabon	4	86	Wood, Manganese, Oil, Uranium
Ghana	4	78	Cocoa, Diamonds, Wood, Manganese
Guatemala*	4	69	Coffee, Cotton, Bananas, Sugar
Guyana	4	83	Sugar, Bauxite, Alumina, Rice
Haiti*	3	68	Coffee, Sugar, Sisal
Honduras*	3	67	Bananas, Coffee, Wood
Iran	1	91	Oil
Iraq	1	92	Oil
Ivory Coast	3	81	Coffee, Cocoa, Wood
Jamaica*	4	75	Alumina, Bauxite, Sugar, Bananas
Libya	1	99	Oil
Malaysia	4	73	Rubber, Tin, Wood, Iron Ore
Mauritania*	1	91	Iron Ore
Nicaragua	5	69	Cotton, Coffee, Meat, Cottonseed, Sugar
Nigeria	3	69	Oil, Peanuts, Coffee
Paraguay	6	77	Meat, Wood, Cotton, Quebracho, Tobacco, Oil seeds
Peru	6	78	Copper, Fishmeal, Cotton, Silver, Lead, Sugar

Table 4.2—cont.

Country	*Number of leading export commodities*	*Export of leading commodities As per cent of total exports*	*Leading export commodities*
Phillippines	3	70	Coconut, Sugar, Wood
Sierra Leone*	3	78	Diamonds, Iron Ore, Palm Kernels
Uganda*	3	83	Coffee, Cotton, Rubber
Uruguay	3	84	Wool, Meat, Hides
Venezuela	2	98	Oil, Iron Ore
Vietnam, South	2	90	Rubber, Rice

* Data for 1966 or latest year for which reports are available.

NOTE: Since these data are based on one year's experience, they should not be used as a final description for any one country. In any one year, the composition of products may shift due to market conditions or internal production difficulties. The purpose of this tabulation is to show the general pattern of dependency on a limited number of products going into the export trade.

SOURCE: Calculated from International Monetary Fund, *International Financial Statistics* (July 1968).

1. Originally reproduced in H. Magdoff, *The Age of Imperialism* (New York: Monthly Review Press, 1969).

Table 4.3

Growth of Manufacturing Production and Employment in Selected Central and South American Countries (yearly growth rates 1950–60)

	Output	*Employment*
Argentine	4·4	2·0
Brazil	9·8	2·6
Chile	5·4	1·7
Peru	6·6	4·4
Columbia	7·6	2·5
Venezuela	13·0	2·1
Mexico	6·5	0·4

SOURCE: Adapted from W. Baer and M. E. Herve, 'Employment and Industrialisation in Developing Countries', in *Third World Employment*, ed. R. Jolly, E. de Kadt, H. Singer and F. Wilson (Harmondsworth: Penguin, 1973).

Table 4.4

Population and Economic Development data in 22 Selected Countries

	Population in millions in 1970	*Per Capita G.N.P. annual growth rates 1960–70*	*Manufacturing sector annual average growth rates*
Trinidad and Tobago	1·0 E	NE	10·0
Panama	1·5	4·2	14·2
Costa Rica	1·7	3·2	9·7
Nicaragua	2·0	2·8	7·6
Singapore	2·0	5·2	14·8
Jordan	2·3	2·9	15·2
Jamaica	2·5	3·5	5·0
Puerto Rico	3·0	—	6·5
Zambia	4·1	7·1	13·8
Sub total	20·1		Average: 10·7
Iraq	9·7	2·5	6·8
Venezuela	10·4	2·3	10·5
Malaysia	10·9	3·1	6·4
Peru	13·6	1·4	7·5
Taiwan	14·0	7·1	16·1
Iran	28·7	5·4	11·2
South Korea	31·8	6·8	16·9
Turkey	35·2	3·9	11·5
Thailand	36·2	4·9	8·7
Philippines	36·9	2·9	8·5
Sub total	227·4		Average: 10·4
Brazil	92·8	2·4	7·8
Mexico	50·7	3·7	7·4
Pakistan (plus Bangladesh)	130·2	2·4	15·0
Sub total	273·7		Average: 10·0

E = Estimated.
NE = No estimate.

SOURCE: Columns 1 and 2 adapted from *Trends in Developing Countries* (World Bank, 1973). Column 3 taken from H. B. Chenery, 'Growth and Structural Change', *Finance and Development*, vol. 8, no. 3 (Sep 1971), pp. 25–26 and is printed in Bill Warren, 'Imperialism and Capitalist Industrialisation', *New Left Review*, 81 (Sep–Oct 1973).

Table 4.5

Direct Company Foreign Investment, United States and United Kingdom, 1929–68[1]

Area	*U.S. capital*						*U.K. capital*			
	1929	*Investment stake 1949*	*1959*	*1968*	*Annual average investment flow 1960–4*	*Annual average investment income 1967–8*	*Investment stake 1960*	*1968*	*Annual average investment flow 1964–9*	*Annual average investment income 1967–9*
total ($ billion)	8	11	30	65	3·5	5·0	12·6	19·7	1·2	2·3
by regions (per cent)[a]										
Europe	19	14·5	16	30	38	22	10	13·7	18·5	13
Canada/United States	25	31	33	33	30	18	24·5	23	19·5	23
Latin America	33	39	35	17	12	27	36·5[c]	30·3[c]	22·5[c]	30[c]
other undeveloped / sterling developed[b]	23	15·5	16	20	20	33	29	33	39·5	34
by sectors (per cent)										
manufacturing	24	33	32	41	50	26	32	36	41·5	28·5
oil	15	29	33	29	20	46	28	25	21	44
mining	16	10	10	8	10	13				
utilities	21	11	9	4	10	15 (utilities and other combined)				
other	24	17	16	18	10		40[d]	39[d]	37·5[d]	27·5[d]

[a] Regional distribution for U.K. investment excludes oil and finance.
[b] U.K. investment in sterling developed areas includes Rhodesia.
[c] U.K. capital figures are combined for Latin America and other undeveloped countries.
[d] U.K. capital figures are combined for mining, utilities and other.

SOURCE: U.S. Department of Commerce (1968, 1970, 1971b); Department of Trade and Industry (1972); Central Statistical Office (1970).

1. Originally reproduced in M. Barratt Brown, *The Economics of Imperialism* (Harmondsworth: Penguin, 1974) pp. 208–9.

Table 4.6

Foreign Direct Investment of U.S. Enterprises in Manufacturing Subsidiaries, by Areas, 1929–69 (book value in millions of dollars)[1]

Year	*All areas*	*Canada*	*Latin America*	*Europe and United Kingdom*	*All other areas*
1929	$ 1,813	$ 819	$ 231	$ 629	$ 133
1936	1,710	799	192	611	108
1940	1,926	943	210	639	133
1950	3,831	1,897	781	932	222
1957	8,009	3,924	1,280	2,195	610
1964	16,861	6,191	2,507	6,547	1,616
1969	29,450	9,389	4,347	12,225	3,489

SOURCES: U.S. Department of Commerce, *U.S. Business Investments in Foreign Countries* (Washington, D.C.: Government Printing Office, 1960) p. 96; and *Survey of Current Business* (various issues).

1. Originally reproduced in Raymond Vernon, *Sovereignty at Bay* (Harmondsworth: Penguin, 1971) p. 71.

Table 4.7

Percentage Increase of Foreign Investment of U.S. Enterprises in Manufacturing Subsidiaries, by Areas 1929–69[1]

1929	Total all areas	1,812
	Europe	629 = 34 (per cent)
	Canada	819 = 45
	Latin America	231 = 12·7
	Other areas	133 = 7·3
1957	Total all areas	8,009
	Europe	2,195 = 27 (per cent)
	Canada	3,924 = 48·9
	Latin America	1,280 = 15·9
	Other areas	610 = 7·6
1964	Total all areas	16,861
	Europe	6,547 = 38 (per cent)
	Canada	6,191 = 36·7
	Latin America	2,507 = 14·8
	Other areas	1,616 = 9·5
1969	Total all areas	29,450
	Europe	12,225 = 41·5 (per cent)
	Canada	9,389 = 31·8
	Latin America	4,347 = 14·7
	Other areas	3,489 = 11·8

1. Calculated on the basis of Table 4.6.

Table 4.8

	Europe	*Canada*	*Latin America*	*All other areas*
Flows of direct investment from the United States	8·1*	6·8	3·8	5·2
Income of this capital transferred to the United States	5·5	5·9	11·3	14·3
Net	+2·6	+0·9	−7·5	−9·1

* Figures in billions of dollars.

SOURCE: H. Magdoff, *The Age of Imperialism* (New York: Monthly Review Press, 1969) p. 198.

5

The Transformation of Indigenous Social Structures under Colonialism

'Our Sorrowful Age'

Now those who were destined one day to possess this country have entered into possession; and they are ruling there, but they do not follow the right path.

They put down the men of worth and exalt the worthless; and if even our lords tremble before them, what of the poor peasants?

They have risen up against God's holy religion; they will not prosper.

(Fulani poem, French Guinea, composed between 1900 and 1910[1])

In the previous chapter we discussed the social, political and, more extensively, the economic degeneration of all those societies everywhere in the world which were brought into the orbit of the advancing capitalist Western system. *Regenerating* forces, however, worked alongside destructive ones, and transformed into a Western mould such aspects of the social order of indigenous societies as would *mediate* and *cement* the newly established economic relationship between the Western metropolis and the conquered peripheries.

This *regeneration* took two main forms:

(1) It involved the reorganisation of the domestic social *hierarchy*, particularly the cultivation of ruling social classes whose interests (economic, social, political and psychological) would be immediately bound up with the systemic operation of the capitalist metropoles, and who therefore could be entrusted to act as deputies for these dominating capitalist centres;

(2) It involved a wider cultural and structural diffusion of values and norms, of beliefs and social institutions which would make the organisation of the colonial societies in its main outline as nearly as possible resemble the organisation of Western societies whose ordering principle – as we observed in the first part of the book – lies in the relative autonomy of the major functional spheres of society, especially the economic.

In this chapter we shall concentrate on the first form of regeneration under colonialism, and we shall discuss the structural problems and characteristics that have arisen from this structural reorganisation under colonialism. In the next chapter we shall discuss the characteristics and problems arising from diffusion.

THE REORGANISATION OF THE DOMESTIC SOCIAL HIERARCHY

It is important to remember that the first contact of the Europeans in the overseas trading areas had often been with societies that were structurally at least as complex and advanced as Europe itself was at that time (feudalism). The mercantile adventures, however, had reduced these complex civilisations in three centuries of European-dominated trade to social formations that had become, by the beginning of the nineteenth century and the onset of colonialism, structurally less complex than the European societies, or indeed than they themselves had been.

Even so there *remained* differences in structural complexity between the conquered territories, and these differences, as we shall see below, had vast implications for their reorganisation under colonialism. Broadly speaking the colonial powers respected, and used, whatever they perceived to be existing viable structures of traditional political authority, whilst they created new structures of leadership and authority, where they felt that the society was too primitive and diffuse to have such differentiated structure.

DUALISM AND INEQUALITY

In *all* instances of colonial penetration, however, the dominance of foreign capital over the export sector created an *over-all dual* structure of economy and society. A dual economy is one in which, to quote Boeke, to whom we are indebted for the term,

> the line of demarcation between big business and the village is sometimes so sharp that the western enterprise appears like a capitalist enclave in a foreign land. A mine or a rubber plantation in the Outer Provinces carries on without touching native life at any point. The capital involved and the labor employed are both imported; the land was waste land, the product is all exported, and even the necessaries of life for the workers have to be brought from elsewhere. The whole concern is detached from its surroundings, although its indirect influence on these surroundings is penetrating.[2]

This view, which described the colonial society of the Dutch East Indies in the early decades of this century, we find echoed much more recently, and with reference to another continent (Latin America), by Celseo Furtado:

> The economic structure of the region into which the capitalist enterprise has penetrated does not necessarily become modified as a result of that penetration . . . displacement of the European frontier almost always resulted in the formation of hybrid economies in which a capitalist nucleus, so to speak, existed in a state of 'peaceful coexistence' with an archaic structure.[3]

The creation of social overhead capital – so frequently claimed to represent a beneficient spillover from foreign domination – in most countries only served to accentuate economic dualism in physical and geographic respects: roads, railway-lines, irrigation projects, port installations, harbours and commercial and administrative cities were built to facilitate the operation of foreign extractive activities. Rarely did these projects become part of the existing internal economic structure: because they were simply built in the wrong places. An illustrative example is that of the telephone connections between major African cities: until very recently they ran via the Paris and London exchanges. The weekly magazine *West Africa* reported that in 1968 out of the 832 inter-African connections 380 still ran via Europe.

Thus capitalism under the aegis of colonialism created *dependent dualist* economies. This structural characteristic continues to be the most conspicuous one in all developing countries. Its logical corollary is *extreme inequality* in the social structure. The productive export sector which is dominated by foreign capital is relatively modern and technologically advanced; productivity is high, as are the incomes of the limited numbers of local workers employed. Modern amenities and services (hospitals, schools, electricity, tap-water, telephones, and so on) are concentrated in this sector which also tends to be relatively urbanised. Western values associated with work, family life, religious practices and social relationships predominate. In short, social dualism accompanies economic dualism. For, outside this modern westernised sector, a large subsistence sector continues to proscribe the lives and careers of the majority of the population. With farming methods still firmly entrenched in traditional techniques and forms of social organisation, a meagre living is scraped off the soil, the surplus – if any – being transferred through trade or taxation to the urbanised modern sector. In colonial times, moreover, peasants were conscripted as labourers and were forced to leave the fields to make roads and to leave their villages to build towns.[4] Thus in many ways the modern sector fed, and today still feeds, parasitically on the traditional sector. The peasants are 'those who pay the bill', as Nyerere said.[5] It is this continuing, dynamic and organic relationship within the national economy between the modern, export-orientated sector, and the traditional domestic sector, which has led Andre Gunder Frank to reject the notion of dualism altogether and to plead for a structural holistic approach.

> Capitalist penetration as well as converting Latin America into a satellite of Europe, rapidly introduced *within* the continent essentially the same metropolis satellite structure which characterized Latin America's relationship with Europe. The mining and mineral export sector was the nerve and substance of the Colonial economy, and though always a satellite to the European metropolis, it became everywhere a domestic metropolitan centre with respect to the remainder of the economy and the society. A series of satellite sectors and regions grew up or were created to supply the mines with timber and fuel, the miners with food and clothing, and the non-working mine owners, merchants, officials, clergy, military personnel, and hangers-on with those means to sustain their parasitic life which

they did not import from the metropolis with the fruits of the indigenous and imported forced labour they commanded.[6]

There is something to be argued for both conceptions, that is of both dualism and structural holism, if we remind ourselves of the diversity of the interpenetration of Western capitalism: whereas in some areas Western capital was accompanied by Western management, that is settlers who undertook *directly* the production of minerals and export crops, and in so doing created capitalist 'enclaves'; in other areas Western capitalism organised the production and marketing of primary products *indirectly* through the use of a chain of middlemen, forestallers and wholesale traders who would in their turn act as 'organisers' of export production in the hinterlands, and in this way incorporate the 'whole of the social fabric' in the capitalist system. In both cases, however, and this is their crucial common feature, whatever *domestic* benefits were derived from the economy were concentrated in newly built administrative and commercial urban centres. Because these centres were unproductive in themselves, their parasitic growth had an unfavourable impact on the countryside causing a reduction in the standards of living of the peasantry. One form of rural protest and reaction against the increasingly wider gulf between town and country has been *rural–urban migration*, resulting in 'exploding' cities. As Hoselitz puts it:

> But the divergent trends in economic development within these cities and outside them, in the wider countryside, had the effect of creating a situation which tended to counteract and eventually turn the parasitic impact of these cities into its opposite. The increasing difference in average income which could be earned in the city as compared with the countryside tended to attract migrants. . . . The population of the city swelled, a laborforce came into existence which served not merely in the trading and domestic service needs of the foreigners, but which made the establishment of industry attractive.[7]

Hoselitz ends on an optimistic note here. The more frequent phenomenon has rather been a distressing discontinuity in the evolutionary process, namely *over-urbanisation*, which by Hoselitz's own criteria means that urbanisation (that is rural–urban migration) in a country is running *ahead* of industrialisation and the development of administrative and other service occupations characteristically concentrated in the cities; and that there is a great disproportion between the costs of urban growth and the maintenance of proper facilities for urban dwellers on the one hand, and the earning capacity of the people congregated in the cities on the other. For example, it has been calculated that an average subsistence budget in an Indian city, that is one which would provide sufficient food and clothing (but little else), requires 250 dollars a year, whereas the typical Indian city resident – at the time of reporting in 1964 – only earned 60 dollars a year.[8] In consequence, the crude death-rate in Indian cities often markedly surpasses that of villages even though the city is more liberally sprinkled with medical facilities.

In this way the peasant has unwittingly tried to redress the balance between the dual sectors of his society. Not, however, to his own benefit. In 1950, the world had 75 cities with a million or more inhabitants, two-thirds of them in the developed world. By 1970, the number of over a million cities was 162, equally distributed between developed and developing regions.[9]

As a consequence of too rapid urban growth and the concomitant lack of housing facilities, so-called squatter or shanty towns have sprung up skirting the cities of underdeveloped countries with the ever-increasing tide of human misery flimsily sheltered in corrugated iron or mud huts, sometimes built overnight. It is estimated that in Latin America the squatter population is as high as 20–5 per cent of total urban population.[10] In these squatter towns (or *barriades* or *favellas* as they are called in Latin America) sanitary facilities are practically non-existent as is access to water. The peasant newcomer to the city often sees these districts as a passage to a better life, a promise not always fulfilled. Recent Latin American studies, for example, show that between 20 and 40 per cent of the adult squatter population were born and bred in these *favellas*.[11] The abject misery of the conditions of life in these settlements have earned the inhabitants the famous label of 'lumpenproletariat'.[12]

COMPRADORISATION

The dominance of foreign capital over the underdeveloped territories also gave rise to merchant capitalism, rather than to industrial capitalism on the part of the local elites, whether existing or emerging. We refer to this social structural process of articulation of the underdeveloped to the developed economy as *compradorisation*. This word derives from the Portuguese word 'comprador', meaning 'interpreter' or 'middleman'. Compradors are people who function as local deputies for foreign capitalists. Since the foreign capitalists were only interested in extracting agricultural surpluses, not in industrialisation, the comprador elite consolidated itself as an *externally dependent merchant capitalist* class, acting as middlemen between the foreign capitalists, say the import–export companies on the coast, and those areas in the hinterland where production itself was not – or today no longer is – directly under the control of foreign companies.

Depending on the level of social evolution that had already obtained prior to European colonisation, two rather different comprador social formations arose: comprador feudalism and comprador tribalism.

Comprador feudisation

In those areas of the world where there had already been some degree of political centralisation, with a readily definable political organisation of rural social relations and a clearly identifiable tax route via feudal fiefs, it was in the interest of the colonial powers to re-affirm and support the position of traditional rural elites, and to ally the interests of these elites firmly to their own.

In order that we may fully understand the social processes involved in this structural transformation, a word needs to be said first about different types of land tenure. A commonly accepted classification of land-tenure systems is one which distinguishes between three forms of differentiation of land*ownership* from land *use*: patrimonial, prebendal and mercantile. Each of these seem to me to be characteristic of a different stage of societal evolution, but all three have in common the fact that some person or groups of persons claim right to the land used by the peasantry. Such a person exercises *domain* over the land, this meaning ultimate ownership or control over the use of a given area.*

In *patrimonial domain*, control of occupancy of land is placed in the hands of lords who inherit the right to domain as members of paramount kinship groups or lineages. This control implies the right to receive tribute from the inhabitants in return for permission to use the land. But permission to use it is not all that is involved. The relationship between lord and peasant originates in a contract in which the lord exchanges *protection* as well as access to the land for the right to receive peasant dues. As an old proverb in fourteenth-century England had it, the peasant promised 'to sweat and sow for both of us', while the lord was to 'keep holy church and myself from wasters and wicked men'.[14] This patrimonial domain over land is perhaps the clearest form of feudalism proper. It was prevalent in Europe throughout mediaeval times.

Prebendal domain, by contrast, is characteristic of a higher stage of societal evolution, for we encounter it in those societies where there is already a high degree of political centralisation. Prebendal domain over land is granted to officials who draw tribute from the peasantry in their capacity as servants of the state. The term 'prebend' refers to grants of income in return for the exercise of a particular office. Prebendal domain is a form of tax collection as it were. The officials who have been granted the domain over a particular piece of land, in turn pass up most of the tribute to the central state as tax. As with patrimonial domain there is some degree of protective paternalism in the relationship with the peasant because the prebendary lords are expected to indulge in certain ceremonial activities to help secure the growth of crops, and to assist with their magic and material aid in cases of flood, drought or other natural disasters.

The big difference between patrimonial and prebendary domain is the aspect of hereditary property rights which is associated with the former, but which is totally absent in the latter. They have in common, however, the inalienable right of the peasant to *use* the land. The land cannot be bought or sold, nor can the peasant be dispossessed.

This latter feature appears only at a later stage of societal evolution with the introduction of the market principle. Typical of this stage is *mercantile domain*, where the land is viewed as *private property* of the landowner to be bought and sold, and used to obtain profit for its owner. The concept of private property turns land itself into a commodity. It is no longer a mere instrument of labour. This

---

* The following discussion on types of domain follows closely Eric Wolf's classification in his classic text on *Peasants*.[13]

thoroughly affects the relationship between lord and peasant since it introduces the possibility of free competitive price formation for the commodity 'land', and henceforth the sum paid for tribute (now called *rent*) by the peasant cultivator becomes, not a portion of the financial yield of the harvest (as it was under both patrimonial and prebendal domain), but a payment made in advance, and in no way dependent upon or connected with what the harvest brings in.

We side-tracked into these different types of rural social structure, each characteristic of a particular stage of societal evolution, because it will give us the theoretical framework needed to understand the impact of colonialism on existing indigenous social structures, particularly in Asia. The colonial government desired to control the natives through what it perceived to be existing traditional authorities, and it needed to collect taxes as well as stimulate the growth of export crops for a world market. The upshot of these various objectives was a policy which, in effect, propped up existing – often decaying – prebendal systems by adding on to them elements of both patrimonial and mercantile domain. In India, for example, the prebendal lords (zamindars) of the decaying Mogul Empire were given tenure (that is *hereditary* property rights) along with the instruction to collect taxes for the colonial state, but these taxes were no longer a percentage of the financial yield of the harvest, rather they were a fixed sum based on the assessment of the value of the land. Moreover, peasants could be ejected for failure of payment, and their dispossessed land could be sold out by those zamindars who failed to reach their tax quota to more ruthless proprietors.[15] This imposed tax scheme was called the permanent settlement scheme. Thus colonial rule imposed a mixture of three forms of social extraction where only one had existed before. Advanced capitalist and retrogressive feudal relations were introduced simultaneously. At the same time the systematic destruction of indigenous handicrafts and arts (through competition from cheap Western manufacturers) prevented the rise of indigenous bourgeois classes, such as had emerged in the evolutionary process in Europe, that could and should have destroyed the power of the landed elites. Rather, colonialism succeeded in firmly entrenching the position of these elites. As argued previously (see pp. 72–3), the collection of taxes under colonial rule involved the orientation of agriculture towards the mono-production of export crops. Small wonder that the interests of the landlords became closely tied to those of the foreign capitalists.

> If security was wanting against extensive popular tumult or revolution, I should say that the Permanent Settlement, though a failure in many other respects and in most important essentials, has this great advantage at least, of having created a vast body of rich landed proprietors deeply interested in the continuance of the British Dominion and having complete command over the mass of the people.[16]

And the iron grip of the zamindars, with but minor moderation, continues until this very day.

In this manner indigenous social evolution became *distorted* under the impact of colonialism. Unlike Europe, where capitalism had superseded feudalism, capitalism and feudalism were made to coexist and to compound the exploitation of the masses under colonialism. As Baran notes,

> A *complete* substitution of capitalist market rationality for the rigidities of feudal and semi-feudal servitude would have represented, in spite of all the pains of transition, an important step in the direction of progress. Yet all that happened was that the age old exploitation of the population of underdeveloped countries by their domestic overlords was freed of the mitigating constraints inherited from the feudal tradition. This superimposition of business mores over ancient oppression by landed gentries resulted in compounded exploitation, more outrageous corruption, and more glaring injustice.[17]

The mitigating constraints of traditional feudal relationships which Baran refers to were, first, the traditional rights of peasant tenants to the use of land (in exchange for peasant dues) under both patrimonial and prebendal domain, and, secondly, the paternalistic protection by the landlords which materialised during times of bad harvests, when the lord would feel obliged to soften tax demands or even feed the peasants from his own well-filled granaries. The introduction of transferable property rights which made for mobility of land as a commodity ended both these traditional rights and these forms of protection.

The marriage between capitalism and feudalism had one more stultifying effect. Enterprising landlords should have used their 'commodity' land as a source of credit, thus accumulating capital to increase agricultural productivity both per head and per acre (that is by intensifying agricultural production through capitalisation). But here an important difference between the social evolutionary processes in Europe – where mercantile domain originated – and the colonial territories where it was imposed revealed itself. In expanding north-western Europe the claimants to mercantile domain had to invest their capital in transforming the productive base of the peasant economy because industrialisation absorbed the 'freed' (read 'dispossessed') peasants, indeed sucked them away from the land. Therefore landlords were forced to turn into capitalist farmers who ran their mercantile domain as a modern capitalist enterprise. This, of course, greatly increased the productivity of the land. But in the colonial territories the fortification of feudal rights, the abundance of a landless proletariat with nowhere to go because of the lack of industrialisation, coupled with the vagaries of the world market and single-crop production, made *extensive* rather than intensive agriculture a natural option for landlords. The marriage between capitalism and feudalism failed to change the productive base. Landlords were content with obtaining as much rent as possible (rent capitalism) and investing it in creating a power base for themselves in the growing commercial and administrative urban centres. Rarely did they engage in rational agricultural enterprise.

A similar fusion between feudalism and capitalism prevailed in Latin America. The only difference, of course, was that they were both introduced there without the co-operation of any existing *indigenous* elites. Rather, early conquests had brought a *settler* landowning class. The Spanish and Portuguese kings had granted the white settlers – originally army officers, or 'conquistadores' – domains with the right to collect taxes, and to carry out judiciary and military functions. In these domains, or *encomiendas* as they were called, the life of the native population was *entrusted* to the care of the conquistadores. However, as we have observed before, for theoretical purposes Latin America after independence can be considered as a semi-colonial state, this time of the American and British, and not the Spanish metropoles. For it is only in that period that the fusion between feudalism and capitalism was achieved. With independence, and under the force of foreign demands, the 'crown' lands became subject to transferable property rights. And from then on the mixture of feudal and mercantile forms of social extraction set in motion a process of underdevelopment in its main outline similar to that described above for India.

Comprador tribalisation

In Africa, the reorganisation of the domestic hierarchy under colonialism proceeded along quite different lines.

Four hundred years of slave trade with the Europeans had liquidated such centralised political states as had existed before, and social evolution had reverted back to primitive and advanced-primitive levels. Social organisation based on and confined to *village communities* predominated throughout the continent, and the few surviving larger slave-based archaic empires such as the Muslim states of Western Sudan, Dahomey, Egypt, Algiers and the political kingdoms of Benin and Yoruba were energetically wiped out by the colonial governments of Britain and France. The latter action, I have no doubt, was taken for the best of humanitarian reasons of the time.

Here, as elsewhere, the colonial powers were anxious to maintain law and order and to raise revenue, through what they thought were traditional authorities. However, structural deformations were wrought, first, because the colonial powers thoroughly misunderstood the function and the role of African chiefs, and, second, because for purposes of administration and economic exploitation the existing village communities were regarded as too small, and were therefore 'regrouped' into districts along what were considered 'tribal' lines.

With due respect to great regional variation, it is yet, I think, permitted to summarise the essence of African chieftaincy as the sacred incarnation of the social solidarity of the community. Land is held in common and entrusted in the 'stool' of which the chief – elected by the elders of the community and from among paramount lineages or descent groups – is the temporal living guardian. The notion of property, or the idea of differentiation between landownership and land use, is totally alien to this system of social organisation. Along with the land, all the values and social institutions of the community are also vested in the chief. That means that political, judiciary and religious functions as well as economic functions – or more correctly, functions of goal attainment, integration, pattern maintenance and adaptation – are diffusely concentrated in the chieftaincy. When the chief delegates his functions he delegates them all simultaneously, to a 'fake chief' or a 'stranger chief' or a 'speaker'. Abuse of power by the chief is prevented by the simple exigency of his removal, a process called 'de-stooling'. In the event of his resisting such attempts, disaffected sections of the community may simply break away and set up their own chiefdom. More importantly, economic, political, judiciary and religious decisions are arrived at by discussion amongst the elders of the community (or adult-initiated members of the community) 'until they agree', which, in practice, means until the chief who has silently sat and listened throughout the discussion gets up and sums up the mood of the meeting.

A group of villages may be loosely connected, though not organised in any strict sense of the word, on the basis of recognised descent from common ancestors. Generally such a group of villages would therefore be said to be of the same ethnic stock (that is 'tribe') and their language would have a great deal in common. But, as Walt Rodney notes, 'beyond that, members of a "tribe" were seldom all members of the same political unit. Very seldom indeed did they all share a common social purpose in terms of activities such as trade and warfare.'[18]

Colonial rule intruded into this system and deflected it from its evolutionary course by making two cardinal inroads.

(1) It perverted the role of the chief by investing in him – both on grounds of expediency and out of naive misapprehensions – far more power than he traditionally possessed. Colonial rule weakened or removed existing checks on chiefs by discouraging the practices of de-stooling and of secessions, and in their negotiations with the chiefs they presumed him to have absolute power and to be the sole decision-maker in the community.[19] The upshot of these policies was that chiefs were made protected stooges of the colonial regime, were alienated from and often hated by their peoples, and that with the undermining of the traditional meaning of chieftaincy the solidarity of the community itself was eroded. This was to have far-reaching consequences for the future domestic social hierarchy: *new* social forces unleashed by the colonial system's introduction of education and its need for a literate westernised class of intermediaries in towns, fostered a *new* political elite which distanced itself from the traditional chiefs and which assumed the responsibility for leading the nationalist struggle for independence. Because of chiefs' collaborations with the colonial regime they lost their power base after independence, at least inasmuch as the formal political structure was concerned. At the same time, however, because of the sanctity of their ancient position they continued to exercise effective authority over the masses of illiterate peasants. In short, the over-all dualism of the social structure which we discussed earlier on was complemented by a dualism in the political structure. This whole process was further compounded by a bifurcation

in the legal system. In order to prop up the position of the chiefs, colonial regimes thought it necessary to re-affirm traditional native law, upon which – they thought – the authority of these chiefs rested. Frequently this even led to a codification of native custom.* At the same time, whilst thus re-asserting traditional native law on the one hand, especially in civil cases (for example, family and marriage customs), colonial governments insisted on submission to their own legal codes whenever and wherever necessary to suit their own economic and administrative interests (which, in practice, meant within the emerging administrative and commercial urban centres), and also whenever native law conflicted appreciably with their own moral codes or, as the French put it, with the 'principles of civilisation'.[20]

Whilst respect for, and the codification of, native customs may indeed have gone some way in bolstering the position of those who administered native law, that is the chiefs and their retainers, the simultaneous operation of two incompatible legal systems at the same time also led to an erosion of these customs and of the traditional system of social control, since individuals could find an escape from it by appealing to Western legal codes, and by moving to the urban centres which resorted under Western systems of litigation. It was not uncommon for colonial governments to distinguish between those who, after registration in the towns, ports, mining settlements or European plantations, were placed under the written statute of the colonial law, and those who remained under native authority. The former depended, as for example the Belgians stipulated in the Congo, on the native's 'evolution' to be affirmed by matriculation.

> Every Congolese had the right to be registered as soon as he had attained his majority according to the Civil Code, if he could show by his training and his way of living that he had reached a stage of civilisation that proved him fit to enjoy the rights and fulfil the duties stipulated in the written legislation.[21]

The upshot of this compounded dualism was the emergence of a new westernised elite, in the pay of colonial administration and exhibiting Western tastes, styles and manners, and antagonistic to the power of traditional elites in the rural areas.

(2) Regardless of large differences between British, French and Belgian colonial rules, all colonial administrations, for economic and bureaucratic reasons, practised a 'district' policy. They would group

---

* This colonial practice, incidentally, was fairly common everywhere. The Dutch codified 'het Adatrecht' (that is native law) in the Dutch East Indies, and even went so far as to encourage aspiring colonial officers to follow 'Adatrecht' as an optional course at Dutch universities. In India, in the 1770s the then Governor General William Hastings ordered the codification and the translation of the laws for the Hindus according to the Brahminical Doctrine as contained in the Dharmastras; and that for the Muslims according to the orthodox interpretation of the Islam. Besides thus encouraging a conservatism of the two legal systems, this policy, of course, also contributed to the mutual antagonism of the two major communities in India.

villages together into larger units called 'districts'. Representing the colonial administration in each district was a district officer, and the need to find him a native counterpart with whom to negotiate, and who would represent the group of villages at this 'invented' district level, led to the institution and appointment of 'paramount chiefs'.[22] Where colonial rule was trying hard to be clever and enlightened it would respect the common descent of the villages in order to make the right kind of amalgamation into districts, and thus, in effect, demarcated more clearly existing but often latent tribal divisions.[23] Codification of native custom for each 'tribe' further deepened these divisions and was in no small way responsible for fostering inter-tribal tension. The legacy of this effective policy of *pluralism*, born from a mixture of administrative expediency of divide and rule, and an enlightened respect for the dignity of indigenous societies, has wrought probably the single most important social obstacle to national development in African states today.

Excepting certain parts of East and South Africa where climatological conditions were favourable to European settlement, and excepting those tracts of land which had to be requisitioned and leased for mineral explorations, colonial governments were quite content to recognise and protect 'native territories' by law, and to respect the system of communal ownership of land, as long as the chiefs would co-operate in the collection of revenue, the conscription of labour for public works and would prod their subjects to grow export crops. A problem of articulation of the economy of these communities with the capitalist world market arose, however, because of the generally primitive level of social organisation of these village communities. The characteristic fusion of religious, social, political and economic forms of behaviour and interaction did not permit utility-rational economic enterprise on the part of the members of the community itself. And since the European capitalists for their part were not prepared to carry on trade beyond the protective environment of the colonial coastal towns, the mercantile link had to be established by a different category of middlemen altogether. What was needed was a group of social aliens, that is of people who had no traditional social connections with the native communities with whom they traded, especially since the nature of the trade was unequal and hence exploitative; these 'pariah' entrepreneurs, as they are sometimes called, were therefore of foreign or socially distinct origin (for example Asians and Lebanese), or members of tribal minorities different from those with whom they traded (for example Ibos in Nigeria), or minorities who were distinct on grounds of their mixed parentage (for example Creoles).*

---

* These pariah entrepreneurs were also common to the other continents. Europeans used social aliens everywhere as middlemen, forestallers and wholesale traders where they had to undertake production of export crops indirectly. Thus, for example, we find communities of Chinese traders throughout South-East Asia in Indonesia, Thailand, Burma, Malaya, Singapore, and North and South Vietnam.[24]

The emergence of these comprador groups complicated further the already plural character of the nation states which were established after independence. As we shall see, barred from formal political power, yet economically the strongest groups, they needed to articulate their interests through the informal channels of bribery and corruption.

6

The Diffusion of European Values and Institutions under Colonialism: Discontinuities in the Evolutionary Process

In the previous chapter I argued that there were two main forms in which colonialism reorganised the conquered territories in order to convert them into proper appendages of the metropoles' societies. But so far we have only discussed one, namely the restructuring of the domestic social hierarchy, the purpose of which was to mediate the new economic relationship between colony and mother country.

Neo-Marxist writers on the subject of underdevelopment, concerned as they are with domination, have tended to focus exclusively on this first form of social transformation, whilst 'modernisation' theorists, being more interested in diffusion, have mainly dealt with the second, that is the transformation of values and social institutions in the colonies so as to bring them into line with those prevailing in the mother country.

Neither bias makes sense. No society can successfully dominate another without the diffusion of its cultural patterns and social institutions, nor can any society successfully diffuse all or most of its cultural patterns and social institutions without some degree of domination.

As repeated many times before, it is the avowed purpose of this book to, if not theoretically combine, at least pragmatically bring together the two perspectives. But such an exercise is an extremely difficult one, since, as domination theorists have rightly pointed out, diffusionist theories, having originated in the West, have themselves been perverted into an ideological cover for domination.

To give a recent, yet classical example: in defending the C.I.A.'s destabilisation activities in Allende's Chile, both Henry Kissinger and President Ford took refuge in a diffusionist-based argument, namely that the U.S. government had the task not only to defend American interests abroad but also to safeguard the freedom of opposition parties in 'democracies' abroad in the very own interests of these countries. The theoretical underpinning of this argument goes right back to the subject matter of the first part of this book, namely, the West's self-reflection and intellectual understanding of its own process of evolution, and the resultant perception of modern society as one where, for example, parliamentary democracy is a logical and necessary corollary of economic development. I am not arguing that either Kissinger or Ford actually and genuinely believe this to be the case, nor do I argue that they do not. Their personal point of view is not relevant. What is relevant is that they could judge such an argument to be socially acceptable to their own public. And indeed, the world's recent history proves that it is. Many thousands of Americans have willingly died, many more Asians have been knowingly killed, not for 'God' or 'country' but in the interest of the 'democracies' in South-East Asia.

And thus we are confronted with a two-level theoretical problem: the diffusionist 'understanding' of modernisation in contemporary developing countries is both true and false.

It is true at the *empirical level* in that the diffusion of western cultural and structural forms has taken place and is taking place in developing countries, bringing in its wake a trail of problems arising out of the conflict of modern and traditional social and cultural patterns: overpopulation, social and personal disorganisation, high rates of violence and crime, increased incidence of witchcraft, religious fanaticism and extremism, corruption and political instability. It is also true that these problems can indeed be 'solved' or 'minimised' when, as a result of skilful social engineering, conflicting traditional patterns are brought into line with modern ones, that is if the modernisation process is speeded up by *comprehensive* social and economic planning as is so often pleaded. Yet the diffusionist approach is 'false' at the *ideological* level, for it fails to understand that its very own theorising aids and abets the continued imperialist domination of the modernising territories by the West. It does so because it draws attention away from the exploitative aspects of a modernisation process which takes place under the aegis of a world-wide *unevenly* developed economic system of capitalist production, and because – as we saw in the Chile example – it provides, however unintentionally, the theoretical justification for brutal forms of political intervention and manipulation by the already advanced modern states, ranging from direct political intervention to the strategic use of food aid.

If we are nevertheless to salvage what is true and hence valuable in the diffusionist approach in order to offset the bias in the neo-Marxist approach, and thereby obtain a fuller picture of the entire range of problems in contemporary developing societies, we must not let ourselves be unduly distracted by the ideological blinkers of the diffusionists, and accept at least by way of working hypotheses their theories and reflections at face value. In other words, we must accept the *bona fide* character of their position and their argument. This holds for the 'theorists', the social anthropologists, functionalists, sociologists and classical economists amongst the diffusionists, but more especially for the practitioners of the diffusionist school, that is the policy-makers, be they U.N. development experts, World Bank advisers of today, or the British and French colonial officials of yesterday.

The following quotations from a British and a French colonial official offer a good starting point for the sort of realistic appraisal that I shall plead for:

In Britain the moral revulsion which prompted the anti-slavery

> campaign found expression later . . . in movements to protect the people of Africa from undesirable exploitation in the very process of their growing contact with the Western World, and finally in the progressive policy developed during the present century. . . . Trade with Africans was not felt by the humanitarians to be a form of exploitation, but just the reverse; they believed that it would help to liberate Africans. . . . Humanitarian aims, diplomatic rivalry, commercial interests – all three contributed to the partition of West Africa by Britain, France, and Germany. But as far as Britain was concerned, it was the anti-slavery movement, the humanitarian factor, which led to the [colonial] involvement from which all the rest followed.[1]

Our French colonial officer is still more blinkered ideologically:

> Take the three great social objectives of French West Africa which obviously relate to the protection of humanity and the dignity of the individual: the freeing of the slaves, education and the fight against epidemics. To replace serf labour by a free peasantry, which means introducing machines and draught animals into the agrarian economy, and to recruit European teachers and doctors and train native ones, money must be raised by loans and taxes. Where is the money to be found? From major export products. But there can be no major export products without plantations and large-scale public works to blast an opening. Trade is indispensable to budgets. This forces the colony to construct harbours and roads, and to discover marketable produce and profitable agricultural methods. *It is constrained to direct natives to wage labour if not compulsory labour.* And thus they become a proletariat. The colony has *proletarianized them in order to free them, educate them, care for them.* In order to carry out social tasks, the colony has had to produce revenue, and to achieve this it has had to become commercial and thus endow its colonials with bourgeois characteristics; they enter the ranks of administration or commerce; they become officials or merchants (emphasis added).[2]

Today, in hindsight, one hardly needs to make a sophisticated Marxist analysis of these statements to discover that the colonial viewpoint was clearly a social consciousness which concealed existing objective relationships by standing them 'on their head' in truly Hegelian fashion. We now know that the colony had to 'free them, educate them, care for them' in order to *proletarianise* them rather than the other way round, and that trade was a new way of enslaving rather than liberating Africans.

The Marxists may score a point here quite easily. But what we should realise is that apart from proletarianising the Africans, the need to free, educate and care for them, precisely because it was so genuine and well-intended, produced a host of other problems of social and personal disorganisation, of overpopulation, of crime and corruption to which Marxists pay scant attention. The problem of overpopulation is a deplorable example. Overpopulation in the colonies, and again after independence, occurred plainly because of the well-intended application of Western medical technology which reduced death-rates whilst traditional patterns of procreation (beliefs about the desirability of large families, a male heir, and so on) continued to keep birth-rates high. There is no other reason for the frightening population explosion in the underdeveloped countries today. Yet domination theorists refuse to recognise 'overpopulation' other than as yet one more symptom of global inequality. Their argument is that in the West birth rates came down as a result of industrialisation, but that in today's world the global capitalist system prevents industrialisation and improvement of the standards of living in Third World countries. Advocacy of family planning and population control for Third World countries, in their view, is but the rich world's answer to a world-wide unequal distribution of production and income which they – the rich countries – are unwilling to alter. This biased form of argumentation was pursued *ad absurdum* during the World Population Conference in Bucharest in 1974 when delegates of Third World countries refused to sign a declaration of intent to make family planning services freely available to all their populations by 1985.

In the opinion of the present author the failure of the Bucharest Conference to get even such a mild resolution accepted is a direct outcome of the dangerous dogmatism and one-sidedness of the neo-Marxist theories of development and underdevelopment. Clearly, a functionalist understanding of the problems of overpopulation, whilst it should stand corrected for its omission of the international-relations aspect of the problem, does hold out a promise of a realistic solution. China, though spearheading the neo-Marxist argument at the Bucharest Conference, herself is an example of a country where birth-rates in recent years have come down, *not* as a result of industrialisation but as a result of successful transformation of cultural beliefs and social patterns. Educational campaigns to bring patterns of marriage, sex and procreation into line with the demographic pressures of medical advance are carried on throughout China and birth-control posters are – next to pictures of Chairman Mao – possibly the most widely distributed posters reaching out to every town and every commune in the land as, incidentally, do the contraceptives.

RIGGS'S THEORY OF THE PRISMATIC SOCIETY

We sidetracked into the population issue because it is an example of a problem of contemporary developing societies, the cause, scope and solution of which is more correctly understood from within the 'functionalist–diffusionist' perspective than from within the neo-Marxist 'domination' perspective. There are more such problems in developing countries today. Problems which issue from the conflict between Western and indigenous modes of organising and integrating the functions of society. To gain a proper theoretical understanding of the nature of these conflicts, reference must be made to the conceptualisation of the evolutionary stages in general societal evolution, as set out in the first chapters of this book. Characteristic of the organisation of modern societies, we observed, is the relative *autonomy* of all the primary functions of society, that is of polity, culture, judiciary and economy, and the high degree of internal differentiation within each of these spheres. The mode of integration of such highly differentiated societies lies in the operation of a formal legal system, based on generalised, universal moral values, and the inclusion of all individuals in the societal community with equal rights before the law. During colonialism this organisation of society was superimposed on the undifferentiated, or, depending on the degree of societal evolution already obtained, less-differentiated indigenous social orders.

An elucidating analysis of the ensuing systemic characteristics and conflicts has been made by Fred Riggs in his book *Administration in Developing Countries.*[3] Adopting the Parsonian model of social evolution as a process of increasing structural differentiation, Riggs labels indigenous, relatively undifferentiated societies, as *fused*, whilst he calls a differentiated modern society *diffracted*. These terms he has taken from the optical sciences. He then invites us to imagine a fused white light passing through a prism, and emerging diffracted upon a screen as a rainbow of different colours. Now, within the prism, there is a moment when the diffraction process starts but remains incomplete (see Figure 6.1).

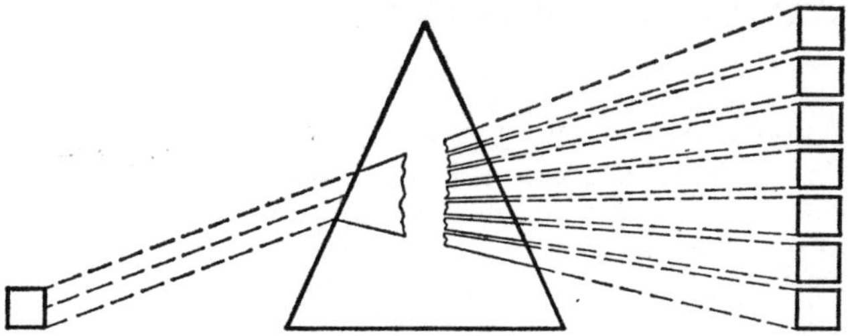

FIGURE 6.1

Using this metaphor, Riggs suggests that contemporary developing societies are typically 'prismatic', that is they present elements of the fused type of societal organisation as well as elements of the diffracted type of societal organisation *simultaneously*. Riggs uses the word 'prismatic' for the sake of this metaphor. It means no more than that contemporary developing countries exhibit a heterogeneity and overlap of these ideal-typical distinct elements. Riggs then proceeds to examine the consequences of this heterogeneity and overlap in each of the main institutional spheres of society, the cultural, the integrative, the political and the economic. In each institutional sphere, these consequences are a 'prismatic' structural quality and a series of what – from a 'diffracted' viewpoint – appear as typical 'problems'.

In the remainder of this chapter I shall use Riggs's *conceptual* framework of the prismatic society in its main outline and purpose. However, in line with the generally eclectic approach adopted in this book, we shall not follow his *theory* of the prismatic society in a rigorous manner. For, athough most of his concepts appear useful in obtaining an understanding of characteristics and problems prevailing in contemporary developing countries – and indeed with the aid of these concepts we shall be able to *extend* the analysis of typical problems beyond Riggs's own analysis – his theoretical generalisations, as is the case with all functionalist theories of modernisation, are deficient. They are deficient because they omit to take into account the historical and contemporary *structural* relations between the developed and the underdeveloped world. This deficiency bears out with special penalty in Riggs's analysis of the 'prismatic' economy, and this is why we shall not treat it in any serious detail. Riggs's approach of the prismatic economy does not go beyond the functionalist observation that both (modern) autonomous market principles and (traditional) social and political considerations are operative in price formation. The result of this prismatic heterogeneity, he argues, is price indeterminacy.[4] Nowhere does he analyse the historically induced lop-sided and ancillary structure of the economy, which would appear to me to be a more correct explanation for price indeterminacy.

THE PRISMATIC CULTURE: POLYNORMATIVISM AND ANOMIE

Typical of the cultural sphere is the coexistence and the conflict between pre-scientific and scientific world-views, each of which provides the individual with radically different orientations for action.

Essential to the scientific world-view, as we saw in Chapter 2, is the objectification of the external world. Nature is seen as 'object', as 'it', subject to mechanical laws, the understanding of which gives man the confidence that purposeful action in accordance with these laws (that is action in terms of cause–effect, means–end) will yield *predictable* results. It is in this sense that modern man's belief system is secular and rational. Or, if one objects to the exclusive application of the term 'rational' for modern man's action, perhaps a better way of phrasing it would be to say that modern man's orientation of action is utility-rational rather than value-rational.

Diametrically in contrast to this view stands primitive and archaic man's world-view which sees the world as explicable only in sacred, supernatural and personal terms. Neither man, nor nature, nor the gods are seen as separable entities. This unitary character of the cosmos is recognised when one says that the world of primitive man is 'sacred'. Nature appears not as 'it', but as 'thou', as a fellow creature endowed with volition, emotion, desires, likes and dislikes just as man himself is. And therefore nature is unpredictable, changeable, *not* subject to mechanical laws. Hence there is no such thing as control over nature. If one wants the heavens to rain, or the trees to grow, or the crops to ripen, one should ask them to do so, and ask them again and again with the return of each season.

The difference in the orientation of action ensuing from each of these contrasting world-views has been succinctly summarised by the

Frankforts: the difference between the 'I' and 'it' relationship on the one hand, and the 'I' and 'thou' relationship on the other, is that, whilst determining the identity of an object, a person is active, manipulative and experimentative, in understanding a 'fellow creature' a person is essentially passive.[5] For subject 'man', the purpose of the relationship, in the latter case, is not to control but to *communicate*. It is this communication with nature that finds its medium in the use of *magic*. Magic usually comprises a rite and a verbal formula, or a physical object (charm) projecting man's desire in the external world with the aim to secure some practical end.[6] Because of the very personal character of the relationship between man and nature, it is thought that long experience, special talents or membership of a specially favoured lineage may improve the 'communication' with nature. Hence, successful rites and incantations become personal 'possessions' jealously guarded against capture by other human beings. In primitive and archaic types of society, each different 'sorcerer' or 'shaman', and each different lineage or clan may have their own magic, their own means of harmonising the relationship with nature.

Although, as we observed in the first part of this book, some societies had reached evolutionary stages well beyond the primitive and archaic ones, it is yet relevant to present the existing indigenous world-views encountered by the Europeans in their conquests in the terms described above. For everywhere in the world, even in societies that had reached the historic stage, the masses of the population remained engulfed in these primitive orientations of action. Indeed, as Parsons has pointed out, it was precisely one of the failures of such historical civilisations as China and India, that *had* achieved the philosophical breakthroughs, to incorporate the masses of illiterate peasants and lower urban classes into the relevant societal community, which stood in the way of the further evolution of these civilisations.[7] The use of magic as a means of harmonising relations with nature, and thus securing practical ends, was – and still is in varying degrees – prevalent in all parts of the developing world.

Mary Kingsley, writing about West Africa, states simply: 'Charms are made for every occupation and desire in life',[8] and Jahoda, speaking about Ghana notes 'the pervasiveness of such beliefs throughout society, up to and including University students. Studies of the clientele of supernatural practitioners indicate that it includes a substantial proportion of literates'.[9]

This latter observation points to the very *coexistence* today of scientific and pre-scientific orientations of action within the same society, and indeed within the same individual. As both Jahoda and Lloyd have argued,[10] instrumental magic is not necessarily used as a substitute for rational effort. A student who is uncertain about his examination results may consult a practitioner and, at the same time, work diligently until the early hours. African peasants, to give another example, customarily are used to secure the growth of their crops by elaborate religious ritual, for example a fertility dance or the casting of a spell. Persuaded by their government's agricultural extension officer they may adopt 'scientifically improved' varieties of seed and fertilisers, but for a long time the latter techniques will not *replace* the former, rather they will be adopted along with the former. It is this coexistence of different orientations of action in the cultural sphere which Riggs refers to when he defines *polynormativism* as a typical characteristic of prismatic societies.[11]

As long as it is physically and psychologically possible for individuals in prismatic societies to adopt both orientations of actions, that is practise magic as well as utility-rationality simultaneously, the coexistence of such competing orientations may not be felt as conflicting nor be anxiety-provoking. In the majority of situations, however, the competing modes of action, and their underlying value premises, cannot be combined in polynormativist symbiosis.

If the reader casts his mind back to our discussion of cultural evolution in Chapter 2, it will be recalled that pre-scientific and scientific orientations of action are intricately bound up with – indeed are a logical extension of – much more fundamental cognitions and beliefs about both physical and human reality. From these cognitions, values are derived which guide not only man's action in the physical world but which also provide the foundations for his social order. And it was these social values particularly which proved impossible to combine in the prismatic situation: universalistic moral precepts which the Europeans brought clashed at every point with the particularistic values of the indigenous peoples; equality before the law was absolutely incompatible with the elevated and sacred status of chiefs, or with the primordial obligations to one's kith and kin; equality of status for women and the associated values of monogamy proved absolutely incompatible with existing polygamous practices; the Christian belief in one 'transcendent god' seemed absolutely incompatible with the worshipping of multiple deities, and so on and so forth.[12] I use the word 'seemed' here on purpose. It is a moot point whether the experience of incompatibility was in every instance spontaneously felt by the recipients of the European 'modern' culture system, or whether in fact the experience of incompatibility itself was induced by the colonial government's predilection to outlaw whichever practices it believed to be contravening the 'principles' of civilisation. Whatever the case may be, where incompatibility occurred, polynormativism gave way to *anomie*, to a turbulent insecurity about norms and values, which built up into moral and emotional tensions that in turn propelled several forms of psychological and social adjustments. Without in any way suggesting this to be an exhaustive list, I have chosen three examples of such psychological and social reactions for closer examination: secret societies, intensified magic and witchcraft, and cults.

SECRET SOCIETIES

In Africa, before the Europeans came, tribal culture and institutions and the socialisation of the young into these (so-called initiation rites) were not 'secret'. They became so, manifestly, in an attempt to protect and preserve indigenous culture and institutions from intrusion by European culture and institutions. However, the *latent* function of the secrecy went well beyond the manifest objective: it effectuated for the individual a rigorous segregation between the clashing western and traditional patterns within the same biographical and social world. From then on individuals could participate in both traditional and modern worlds without fear of social exposure to their contradictions.

Such effective social segregation aided the separation within the individual's own mind of these contradictory beliefs and practices. Thus psychological compartmentalisation of conflicting actions and their underlying values found a social structural outlet in the formation of these secret societies. To date, it is quite possible and psychologically manageable for the prismatic individual in African states, for example, to participate in a Christian church service on a Sunday, and to engage in a ritual sacrifice in the secret society in the 'bush' on the following Wednesday, even though his own friends and enemies may have participated with him on both occasions. Yet individual mental conflict is minimised because the individual knows that social exposure, and therefore the effective operation of social control, is foiled by secrecy. The bush will not and will never reveal its secrets.[13]

INTENSIFIED MAGIC AND WITCHCRAFT BELIEFS

If, for obvious reasons, information on secret societies is scant, the literature on magic and witchcraft beliefs is more than adequate. Anthropologists working in developing areas under colonialism and after independence have often noted an *increased incidence* of magic and witchcraft (so-called 'black magic') beliefs, when as a result of westernisation the impacts of rapid social change make themselves more widely felt.

The belief in magic and the fear of witchcraft are both expressions of the same belief system. Both are pre-scientific attempts to understand reality. But whereas magic is used to communicate with the external world in the pursuit of some practical end and thus is mostly thought to be productive, witchcraft is generally considered a destructive act, and therefore the *belief* in witchcraft may help an individual to explain his misfortunes and failures. For the essence of the belief in witchcraft is that people believe that the blame for some of their sufferings rests upon a peculiar evil power, embodied in certain individuals in their midst.[14] Thus both magic and witchcraft beliefs are designed to help the individual cope in a hazardous and unpredictable world, and it follows that in times of rapid social transformation, such as those occurring since colonialism and westernisation, when unfamiliar and contradictory social situations and events proliferate, the individual will need to resort with even greater frequency to such means of securing psychological comfort as traditional culture makes available to him. This hypothesis, indeed, has been substantiated by anthropological findings in *all* parts of the world.[15]

In Africa today supernatural practitioners proliferate at a truly amazing rate, a reflection of the highly lucrative nature of their business. The consultation of 'Alphamen', 'Muslim men', 'witch doctors', and so on, is commonplace in most African states amongst all ranks of the population – with highly placed politicians and leaders not in the least excepted – and for *all* practical purposes ranging from the procurement of a house, employment or careers promotion, to the obtainment of positive examination results or votes in the ballot box.

Migration to towns and the stresses associated with urban life in a too rapidly urbanising country again greatly increase the number of anomic situations confronting the individual. Redfield, who studied processes of social transition in Central America, found as a corollary of our general hypothesis that beliefs in magic and protective charms against witchcraft are much more frequent in the towns than in the rural areas.[16]

Whereas the belief in magic is, on the whole, quite harmless if unrealistic, the belief in witchcraft is potentially more dangerous, for in times of social stress witchcraft *prosecutions* may become the weapon with which the threatened community defends its values and closes its ranks against deviants and defectors.

To obtain a full picture of the entire range of tension-managing functions of the witchcraft belief, it is advisable to analyse this phenomenon at three distinct levels: psychological, cultural and social structural.

Psychological

Witchcraft beliefs, in this respect, have a cathartic function, either serving as an opiate for the individual, or as a catalyst for his aggression.

The opportunity to blame someone else for one's suffering, and someone not entirely human at that (since, possessed by evil powers, the witch is not considered entirely human) of course offers great psychological comfort, especially since traditional society equally equips one with a range of antidotes: one may swallow medicines, wear amulets or charms, or visit anti-witchcraft shrines. But not only is antidoting against witchcraft a psychological means for the individual to cope with the tensions of change, actual witchcraft prosecutions may serve as a major instrument for dealing with aggression in a society which does not normally permit the use of direct aggressive acts between members. Kluckhohn's analysis of the Navaho Indians in North America confirms that not just antidoting, but also actual trials of accused witches, increased in times of social upheaval and disequilibrium.[17]

Cultural

Here the witchcraft myth serves, as Philip Mayer has pointed out, as a cosmological device. By firmly and resolutely placing the responsibility for human suffering and evil right *outside* its own moral order, a society can keep that same moral order intact. By blaming the witches for all evil and misfortunes, the spirits of the ancestors can remain good and benevolent.[18] Also, by easy resort to witchcraft beliefs, a society can continue to have faith in the expediency of its norms, its rules of practical conduct, and its techniques for survival, even in the face of gross empirical contradictions. Farming methods that yield poor harvests, medical practices that bring no cure, or even established procedures for childbirth that result in high numbers of stillborn, may not be doubted for the witch is always there to receive the blame.

Social structural

The witchcraft belief, in this respect, may help sustain the social organisation of a community by providing what Marwick has labelled a 'social strain gauge' for those structural relations that are under particular duress.[19] Conflicts arising between incumbents of roles which, because of changed historical conditions, are no longer viable, are expressed in witchcraft accusations, thus leaving the structural organisation *itself* uncriticised. We may obtain an insight into this tension-management function of the witchcraft myth by uncovering the relationship between the witch and his/her accuser. For instance, in African polygamous communities, witchcraft accusations today may occur more frequently between co-wives as changes in economic conditions and diffusion from other cultures seem to press for a transition to monogamous types of households. In the same way, matrilineage has become a likely source of tension, and hence of witchcraft accusations. Conflicts arise over property and headmanship now that a man's loyalty has become divided between his own children and those of his sister. Nadel found witchcraft accusations commonly occurring between a man and his maternal uncle amongst the Mesakin in southern Sudan,[20] and Marwick reported a similar finding from his fieldwork amongst the Cewa.[21]

On another continent, in India, too, the witchcraft myth has been recorded to fulfil this tension-management function. Scarlett Epstein discovered an increased incidence of witchcraft accusations in a Mysore village, as a result of women having become money-lenders, which rendered a conflict between traditional values and those of the new economic system.[22]

Philip Mayer sums up the social structural function of the witchcraft myth rather neatly when he says that there is not much difference between the statement 'X hates me' which is the core of the witchcraft accusation, and the statement 'I hate X' which is what the incumbents of an outdated role relationship really would like to say.[23]

The relationship between the accuser and the accused covers only one aspect of the social function of the witchcraft belief. It is also useful to examine the type of character or personality that a society most frequently selects for its witchcraft accusations, and to examine the concrete terms in which the accusations are made. More often than not one will find that the accused person is in some way a social deviant. Thus we learn which norms and values of a society are particularly under pressure. A case in point is the witchcraft trials in Europe and North America during the sixteenth and seventeenth centuries. In those turbulent years after the Reformation witchcraft accusations were couched in terms of accusations of heresy: 'witchcraft grows with heresy, heresy with witchcraft', the Lutheran pastors in Germany used to say. In the witch craze that followed, Catholics and

Protestants alike used the witch mythology to denounce the heretics amongst their own flock. A useful inquiry into the European witch-craze has been made by the historian Trevor Roper. He concludes that the revival of the witch craze in the 1560s was not the product either of Protestantism or Catholicism, but of both, or, rather, of their conflict.[24] And this is what I understand Philip Mayer sees as the essence of the witch belief, namely that it has the function of *affirming social solidarity*, because it searches for and destroys the enemy from within, that is the traitor within the gates.

> The witch is conceived as a person within one's own local community, and often even within one's own household. All human societies require a basic loyalty between the members of the small co-operative and defensive group. The local community, the family, the household, all in one way or another make this demand of loyalty as a categorical imperative. Persons who stand in these intimate relations must on the whole work together, not against one another, if the group is going to survive as a group.... The witch is conceived as a person who withholds this elementary loyalty and secretly pursues opposed interests.... Thus the witch is the hidden enemy within the gate.[25]

Thus seen, witch beliefs are of all times and of all places, and we can compare sociologically African witch beliefs with the post-medieval European witch craze, as well as with McCarthyism, Powellism, and the Communist Party purges of the modern era. Enoch Powell unwittingly yet predictably, in a speech on 18 November 1972, literally spoke of the 'traitor within our gates'. We also now understand why witch-hunting in one form or another increases in times of social change: 'Witch-hunting, then, goes together with a feeling that basic sentiment, values and interests are being endangered.'[26]

In traditional societies it was quite feasible for witchcraft accusations to have a tension-managing function in resolving the social tensions of a community, for, in the past, every witchcraft accusation was taken seriously and every allegation was proved or disproved in the eyes of the community by the method of the poison ordeal. If the poison ordeal proved the accused to be guilty he would indeed be persecuted, but if he was proven innocent then the person who made the allegation was punished equally severely. In this manner poison ordeals contained the witch belief within manageable proportions, for only in extreme cases would one dare accuse one's fellows of witchcraft.

Today, however, these traditional ways of dealing with witchcraft accusations are prohibited by the modern 'diffracted' governments. The result is countless and unresolved 'private little witch-hunts', as well as a premium on antidoting and all other forms of protection against witchcraft. Field, reporting from a particular area of the Ashanti in Ghana, notes that of the twenty-nine shrines located within one small area, twenty-one were less than ten years old. All shrines were privately owned and profit-making.[27]

The same may be said of the many movements led by witch-finders and witch-hunters that have especially mushroomed in the turbulent years since independence. Wonder-prophets and miracle healers, currently tour African countries to remove the practice of witchcraft, a highly lucrative business which itself also of course involves a great deal of divine relations and magic.

Again, the rapid profusion, and the increasing popularity of new sects and 'syncretic' or 'nativist' churches may be accounted for in similar terms. The new sects accept the God of Christianity but specialise in healing and in counteracting witchcraft through the medium of spirit-possessed priests. Lloyd reported in 1955 seventeen different such sects in Accra,[28] but, on visiting that town in 1971, I found the number had increased to no less than 300.

CULTS

This leads us directly to a discussion of a third psychological and social form of adjustment to anomie – the 'religious cult'.

In some respects the cults and the so-called 'nativist churches' in colonial and ex-colonial countries may be grouped together as they were both, in the first instance, *syncretic* attempts to combine elements of the taught Western Christian religion with elements of the more diffuse religious experiences of the natives. The difference between cults and nativist churches is, in my opinion, one of degree of 'enthusiasm', of frenzy and hysteria, which is commonly only associated with the term 'cult'.[29]

Cults have been called 'religions of the lower orders', for they are prone to emerge amongst people who feel themselves to be oppressed and who long for deliverance.[30] Small wonder they flourished amongst the populations of the colonial countries. Beyond the ostensible aim to combine the Western religion with indigenous religions, cults in the *colonial* period served important political and social functions. Under colonial oppression cults served as an *integrative device* uniting segmented social units (for example the disparate villages and tribes) by re-asserting their common ethical values in contrast to the European values, and by re-establishing their own dignity in defiance of the white man's racism. The cults' 'millenarian' dreams of deliverance from white oppression frequently led to – albeit irrational – forms of protest. Expecting and hoping for the 'millenium', cult members would destroy their crops, their cattle and their means of livelihood. A world-famous example are the Melanesian *cargo cults*, so-called because of the people's firm expectation of the arrival of cargo vessels full of the white man's material goods, but without the white man himself. Thus the main *effect*, if not the main manifest purpose, of the millenarian cults in the colonial era was 'to express social and moral solidarity and independence in a highly charged emotional situation resulting from the overthrow or questioning of ancient ethical values.'[31] And, in so doing, the cults welded together previously disparate and hostile groups into some form of political unity against the colonial oppressor. In colonial countries many nationalist independence struggles were born from the cults. Peter Worsley in his excellent book on the cults, *The Trumpet Shall Sound*, makes a very pertinent observation, namely that the general trend in the development of the cults is away from apocalyptic mysticism and towards secular political organisation.

> When secular political organisation has replaced millenarianism, the cults which persist into the era of secular politics almost invariably lose their drive. The revolutionary energy is drained from them; they become passive. The Day of the millenium is pushed farther back into the remote future. The kingdom of the Lord is to come, not on this earth, but in the next world; and the faithful are to gain entrance to it not by fighting for it in the here-and-now with their strong right arms but by leading quiet and virtuous lives.[32]

This observation is particularly relevant in the context of contemporary developing countries. Having lost their political function to the formal political arena, yet with the same anomic ills still unresolved, and if anything increased, cults have now slipped into a more differentiated and specialised function characteristic of religions in 'diffracted' societies. Cults have watered down to 'nativist churches' concentrating more exclusively on the spiritual well-being of their clients. Though they still practise healing, and still counteract witchcraft, and in that sense are more diffuse than our own religions, by absolving their political function they have also ceased to act as a form of social protest. Quite the contrary: by providing psychological security for the individual in these times of particularly severely felt radical change, insecurity and anxiety; by offering irrational explanations for personal frustrations which are – indeed – not rationally explicable at the individual level, since these frustrations are in part generated by extreme internal inequalities and a stagnating economy, the nativist churches and sects *deflect energies* away from rational social analysis and revolutionary protest.

With independence the anomic ills which prompted the rise of cults have not disappeared. And, therefore, in the apolitical form of churches and sects the 'cults' continue to mushroom. No longer recognising the need to unite against a clearly identifiable oppressor, the cults have fissioned into many smaller congregations, again divided and often mutually hostile, and reflecting deep tribal and clan divisions within the society.

THE PRISMATIC SOCIETAL COMMUNITY: POLYCOMMUNALISM; CLECTS; CORRUPTION

The word 'society' in the precise meaning which – following Parsons – we have given this term (see Chapter 2, above) refers to a system of human interaction characterised by common definition of membership, grounded in a common value orientation, and exhibiting a relatively high degree of self-sufficiency.

European conquests, for reasons of imperial rivalries, administrative expediency or missionary competition, created colonial *countries* which were mere geographical expressions of and in no way coextensive with existing 'societies'. Frequently these geographical expressions comprised several 'societies', that is several existing self-sufficient tribal or village communities; in other cases, the boundaries of these geographical expressions cut right across the territorial boundaries of real, existing 'societies'. After independence the emerging nation states were, in fact, these same geographical expressions, this time wrapped up in the myth of national sovereignty and international recognition.

If we are therefore compelled to characterise the nature of the 'integrative' structure, that is of the societal community, of the contemporary developing countries, we have to recognise that most of these countries are comprised not of one but of many 'societies'. In order to avoid terminological confusion, students of developing countries have preferred to use the word 'communities' rather than 'societies' in connection with the real existing societal groupings within each developing country. Thus, many authors have characterised the society of developing countries as *polycommunal.*[33] Sometimes they would speak of 'plural society',[34] to indicate the existence of what on all relevant sociological accounts are several existing societies within one country.

The existence of several societies within one country, however, is but one reason for the use of the term 'polycommunal' in connection with developing countries today. There is another related reason: an important feature of fully differentiated societies is that they are 'mass' societies, meaning that all the members of the societal community (however large) can participate in the dominant cultural forms of the society because of the existence of literacy, radio, newspapers, and physical forms of transportation and mobility. In other words, every member of the societal community can be mobilised for mass communications. In the 'fused' system of societies, on the other hand, 'mass media' do not exist, and hence the population is obviously not mobilised for mass communications; they remain scattered and isolated in small communities.[35]

As Fred Riggs persuades us, a logical intermediary stage is the prismatic society, in which

> The introduction of mass media and widespread fundamental education causes partial mobilization of the population. The rate of mass assimilation to the symbol system of the elites, however, tends to be slower than the rate of mobilization. Hence several large communities tend to arise within the society. The elite of one community is dominant and the counter elites of the 'differentiated' or 'deviant' communities, lacking access to the dominant elite, engage in various forms and degrees of hostile action, ranging from apathy and non-cooperation, through bribery and sabotage to open violence and revolution.[36]

In defining the structural organisation of these communities in the polycommunal country, Riggs refers to three of Parsons's pattern variables which are especially important in describing polar differences of structural organisation between diffuse and differentiated societies, or, as Riggs calls them, 'fused' and 'diffracted' societies. These three pattern variables are:

(*a*) functionally diffuse versus functionally specific;
(*b*) particularism versus universalism; and
(*c*) ascription versus achievement.

Riggs's model of the prismatic intermediary form of social organisation conceptualises the logical intermediaries of each of these three pattern variables.

(*a*) *Functionally diffuse versus functionally specific*

The reader may recall that this pair of pattern alternatives refers to the scope of activities in any one role structure; a functionally diffuse structure being one where religious, economic, social and political activities are not clearly distinct and are carried out simultaneously within the same interactive context, and a fuctionally specific role structure being one where the scope of the role relationship is limited and precisely formulated. This dichotomy reflects most nearly the conventional sociological distinction between *Gemeinschaft* and *Gesellschaft* (Tonnies), between 'community' and 'organisation' (MacIver and Page), and between 'primary' and 'secondary' groups (Cooley).

In the 'prismatic' situation, Riggs argues, we typically encounter social structures whose functions are more diffuse than those of a modern organisation, but at the same time more specific than those of a traditional village community or an extended family group. Such structures, or social groupings, Riggs therefore labels *polyfunctional.*

Examples of such polyfunctional structures are the medieval guilds, and – in contemporary developing countries – the social organisations of 'alien' communities, such as, for example, the Chinese Chamber of Commerce in Indonesia, the Indian organisations in East Africa, or the Lebanese organisations in West Africa, as well as the various urban tribal associations. These polyfunctional groupings, although manifestly set up to promote, say, the economic interest of their members, cater for a much wider range of the individual's needs: educational, recreational, social integrative, and religious.

The evolutionary history, the functions and the scope of especially the *urban tribal associations* in Africa, have been very thoroughly studied and fully documented.[37] Mostly, these studies were carried out in the context of more general studies on urbanisation and the problems of rural–urban migration in Africa. Authors have generally agreed that, by being poly-functional, the ethnic associations have helped to bridge the gap between the traditional 'fused' and the modern complex worlds. For, essentially, the associations are seen to provide two basic needs of the transitional individual: they give him *security*, and they *socialise* him into the ways of town life.[38] They provide security by helping the individual retain his identity and sense of belonging, through frequent contacts with his own tribesmen, by reaffirming the values of his traditional rural home, by disseminating news from home and by organising traditional forms of recreation. At the same time these ethnic associations socialise the newcomer into urban patterns of behaviour. They coach him through the maze of atomised urban social relations – they help him find a job, a house, or they may set him up with a loan (many associations run a kind of savings club) and they may teach him budgeting and accounting skills. Most importantly, the associations provide necessary welfare services which the newly mushrooming cities are still lacking. For example, the associations may care for the sick and the unemployed; they may arrange and help pay for the funerals of the deceased and the repatriation of the descendants.

In other developing countries, the role of such polyfunctional voluntary associations has been less fully documented, perhaps because they have featured less prominently in the transitional situation. Yet they are certainly not entirely absent amongst the migrating populations of Latin America, as some contributions in Mangin's *Peasants in Cities* testify.[39] And, where they are operative, sociologists and anthropologists have attributed them with largely the same functions as those in Africa. The difference – it has been suggested – appears to be that, in Latin America, the voluntary associations 'are more numerous and instead of tribal groupings, in Peru we find migrants forming regional associations which unofficially represent specific places of origin'.[40] Elsewhere, in India, similar functions appear to be carried out by existing traditional caste organisations.[41]

In welding traditional and modern worlds together, and in easing the great transformation from an agricultural to an urban industrial state, the polyfunctional associations have proved immensely valuable. But, as always in human life, for every success there is a price to pay. Some tribal associations, whilst successful in affirming the traditional or regional identity of their members, have stood in the way of the creation of wider national communities. Whilst mitigating the effects of too-rapid social transition, they have frequently stopped the process of social transition in its track, so to speak. As we shall see in further examples, a too-properly functioning intermediate structure may help consolidate what is in fact an unbalanced and distorted over-all societal structure, and thus hinder its further evolution and progress.

(*b*) *Particularism versus universalism*

A second relevant pattern variable characterising the type of social structure in any one of the two societal extremes (that is 'fused' and 'diffracted') is the distinction between particularistic and universalistic application of norms. To recall (Chapter 3, p. 55) particularistic norms are those which include in the definition of the code the social objects to which the behaviour is applicable, for example in the traditional Chinese norm of *filial piety*. Universalistic norms are those which set ethical standards irrespective of who is interacting with whom; in other words irrespective of the social context, for example 'love thy neighbour' or 'maximise benefits over costs'.

This pair of alternatives, again, is dichotomous. It fits the fused and the diffracted models well, since the prescriptive codes of primitive communities are clearly particularistic, whilst the formal legal codes of modern societies are clearly universalistic. If we visualise an intermediate type of norm application, we may characterise our prismatic group. Riggs suggests that the intermediate type of norm application is *selectivist*: 'A norm is applied selectivistically when it is invoked particularistically to discriminate against all members of the community, but universalistically in so far as it makes no discrimination of persons within this community.'[42]

A better way of defining selectivism – it seems to me – is by putting it the other way round: selectivist application of norms denotes a social situation where universalistic moral standards are applied particularistically (that is 'only' to members of a defined social in-group). The classic examples of selectivism are all forms of racial, tribal or ethnic discrimination. I remember well the difficulties which some western medical friends of mine encountered in a hospital in West Africa, where they tried to teach the local nursing staff to apply the ethical standards and requirements of the nursing profession universalistically to all patients regardless of these patients' ethnic backgrounds. Unsuccessful, the doctors finally decided to entrust even medical aspects of the hospital treatment to members of the patient's family.[43]

(c) *Ascription versus achievement*

Finally, a third pattern variable relating to the ideal-typical structural differences between fused and diffracted societies is the pattern alternative: ascription versus achievement, as a principle of role, resource and status allocation.

In primitive 'fused' communities, social resources of power, wealth and prestige are typically assigned on the basis of ascriptive criteria (for example sex, age, descent, membership of paramount lineage, and so forth) whilst in modern societies such social resources are typically allocated on the basis of achievement. Furthermore, the differentiation in the institutional spheres of society, in conjunction with the achievement principle of stratification, makes it possible for individuals to move up or down in the social hierarchy in any one sphere (say in the economy) without having to move simultaneously in all other spheres as well.[44] By contrast, in 'fused' societies, there is typically an *agglutination* of social values such that the most powerful are often the most wealthy as well as the most prestigious.

The logical intermediary between ascription and achievement in the prismatic situation, Riggs proposes, is again a simultaneous operation of both principles, that is of ascription and achievement, such that the ascribed element takes the form of limiting candidacy for elite status to members of certain ethnic, racial, religious or linguistic groups, or any other sub-group within the society, whilst achievement criteria would apply to the allocation of further positions within the elite group. Riggs suggests the term *attainment* to characterise this prismatic principle of stratification.[45]

With the help of these three 'intermediate' types of pattern variables – *polyfunctionalism*, *selectivism* and *attainment* – we are now ready to characterise the prevailing type of social organisation in contemporary developing countries. With amusing semantic wit, Riggs proposes the term 'clect'. The word is a semantic assemblage of 'cliques' and 'sects', 'which has the added advantage that it already occurs in nature, so to speak, in "eclectic".'[46] Clects, therefore, are types of social organisation which eclectically combine structural elements of both fused communities and diffracted organisations.

The clectic social organisation is the prismatic organisation *par excellence*, and as such it forms a natural counterpart to the *poly-communal* composition of contemporary developing countries.

> Each clect draws its membership from a particular community; it applies its norms selectivistically to members of that community; and its polyfunctional goals always include a communal orientation as well as whatever economic, religious, political, educational, or social objectives constitute its manifest functions.[47]

In this manner, clects both reflect and deepen existing social cleavages within the society, and hamper its progress by arresting the evolution of the integrative function of society towards a definition of societal community which would be coextensive with the nation state, and which would be in keeping with the requirements of modernisation.

CORRUPTION

If the clect is the natural counterpart of polycommunalism in the prismatic society, corruption is the evil companion of both. For, in developing countries, corruption is first of all the vehicle for negotiation between badly integrated structures of the social order.

The fact that I wish to examine corruption as a typical problem of developing countries should not be taken to mean that I believe corruption to be absent in the advanced countries. Few would want to make such an assertion in the wake of the U.S. Watergate scandal, and the British Poulson affair, to mention but two of the headline-hitting examples of recent years. It *is* my contention, though, that corrupt practices in contemporary developing societies are much more pervasive, much more of an everyday pattern of life, and that they disrupt economic life to a far greater degree than is the case in the advanced countries.

Gunnar Myrdal in his recent book, *The Challenge of World Poverty*, though writing on South Asia, emphasises the common ill of corruption in all developing countries, and he severely criticises Western development experts for their 'diplomatic' neglect of the problem of corruption. He insists that the 'taboo' should be broken.[48] This is precisely what I am hoping to do in the following pages.

Sympathetic observers of developing countries are in the habit of pointing out that the very notion of corruption contains a legal criterion which itself has only emerged in the West with the full development of the modern state and the consolidation of the modern rational bureaucracy. They argue that dictionary definitions of corruption as 'the illicit gain of money and employment' or as 'the use of public resources to further private interests' assume a differentiation of political and administrative functions, and of public and private funds, characteristic only of the modern bureaucratic state. This, it is claimed, is the reason why the same practices which were carried on in Europe for several hundred years, and which were only painstakingly and slowly removed in the latter half of the nineteenth century, stir up so much moral indignation and attract so much attention in contemporary developing countries. Such is, for example, the position taken by James Scott in one of his earlier papers on corruption:

> The legal standards now in force in developing nations offer the civil servant or politician much less scope for manoeuvre than his counterpart in Europe a century before. Application of the legal criterion often means calling the receipt of 'side-payments' or government supply contracts in Rangoon in 1965 'corruption' whilst withholding that label from quite similar practices in nineteenth century England.[49]

Undoubtedly, such observation is correct. Yet this labelling theory does not help us at all in explaining the *objective* practice of corruption, and it is this explanation which the very governments of developing nations, because they *have* accepted the modern legal criterion, are urgently seeking. Corruption in contemporary developing countries has become the mainspring of a stream of commissions of enquiry, a focal point for popular discontent, and a near over-used rationalisation for military takeovers.

Here I will argue that widespread corruption occurs everywhere where the organisation of social life is undergoing a transformation from what Riggs has called a relatively 'fused' state to a more 'diffracted' state of society. More precisely, corruption accompanies the discontinuities in later stages of social evolution when economic differentiation, with its attendant integrative mechanisms of money and markets, gets into full swing. The role of money is important: for only with the introduction of money can bribery become an attractive alternative to physical coercion as a means of influence.

The discontinuities in this evolutionary process may be caused either because of the *rapidity* of the process, as for example during the Industrial Revolution in Europe and the United States, or because of the *superimposition* of advanced (that is more 'diffracted' or more 'differentiated') forms of social organisation on lesser advanced (that is more 'fused') forms of social organisation, as has typically been the case with contemporary developing societies during colonialism and after. Thus, although essentially the same corrupt practices occurred in the early and later developers, the *roots* of corruption are different, and it is this difference that is largely responsible for what I believe to be the greater rate and scope of corruption in contemporary developing societies.

With respect to the scope of corruption in developing societies, it is as Ronald Wraith and Edgar Simkins, writing about West Africa, state quite simply: 'In general there is little that cannot be bought and little extortionable that is not extorted.'[50] Corruption sweeps across every sphere of life and affects everyone: with patients offering bribes to nurses in hospital to persuade them to pass on a bed-pan; traffic offenders bribing police officers to waive the fine; tax collectors adding their personal increment to the inland-revenue exactions; councillors awarding contracts to firms in which they (or their kin) have a financial stake; educational officers giving government scholarships to their cousins; and political candidates buying the votes of entire electoral districts. This impressive list immediately shows us one thing, namely that there are several kinds of corrupt practices, and, therefore, that a proper theory of corruption ought to distinguish between them.

But first let us define corruption properly. For a sociological analysis cannot settle for a layman's definition of the phenomenon as 'the illicit gain of employment or money'. Rather we will have to define corruption within the theoretical framework to which we have committed ourselves throughout this book.

If one recalls the discussion on the heterogeneity and overlap of ascriptive and achievement principles of social stratification (p. 126) it will be remembered that there is a difference in the way in which roles are assigned and rewards are distributed in respectively small, homogeneous 'fused' social systems on the one hand, and in large, heterogeneous 'differentiated' or 'diffracted' social systems on the other. We had observed that in the fused system, because of the fusion of functions and hence of social roles, there was also an agglutination of social rewards, that is elites in those societies typically command all three societal resources of political power, economic wealth and social prestige. At the other pole of the evolutionary spectrum, however, the differentiation of societal functions generates a separation of social roles, and hence of elites, and to this in turn corresponds a differentiation of societal rewards, with political power typically going to political elites, prestige and fame typically blessing religious leaders and intellectual elites, administrative roles being rewarded with a fixed salary and security, and economic roles being allocated economic wealth.

Viewed within this sociological perspective, 'corruption' in its widest sense can be regarded as a *process of trade in societal rewards*. Thus, it is a process whereby, typically, power is traded for wealth, wealth for prestige, prestige for wealth, and wealth again for power. And the need for such trade is rooted in the 'prismatic' situation where there are discontinuities in the evolutionary process due to an overlap or heterogeneity of elements of fused social systems and those of diffracted social systems.

Of course, there are different *levels* at which the trade takes place, and there are also differences in motivation between those who accept bribes and those who offer bribes (that is between those who trade power for wealth and those who trade wealth for power). A proper theory of corruption would have to render all these differences intelligible.

Even when one restricts oneself to corrupt practices which involve government and administration (that is political power) four logical categories of corrupt practices, and hence potentially corrupting social groups, suggest themselves.

First we may distinguish forms of corruption at the law-*making* level from those at the law-*implementation* level. The former is 'high-level' corruption, involving politicians, party officials and the higher echelons of the civil service. The latter, corruption at the law-enforcement level, involves the lower echelons of the civil and public services. It is at this level, where contacts between administration and the general public are most frequent, that the greatest *volume* of corruption occurs, although the amount of damage done and money involved may well be greater at higher levels.

A second logical distinction we can make is between the accepting (or asking for) and the offering of bribes.

A combination of these two axes yields four logical possibilities of corrupting pressures, practices and agents, as can be seen from Figure 6.2.

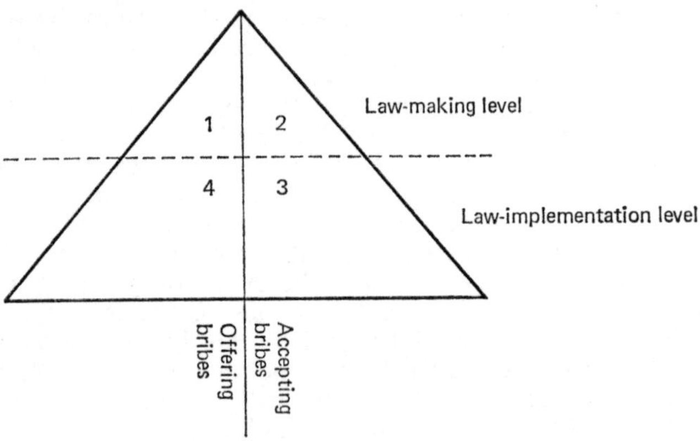

FIGURE 6.2

Let us next examine each of these four forms of corruption in turn, tracing their roots to the prismatic situation.

(1) *Bribery at the law-making level: informal political influence by marginal economic elites.* Two instructive examples, presented in Scott's aforementioned article, demonstrate exactly what this kind of corruption is all about, and how in comparable form it occurs in different times and places.

> In England throughout the seventeenth and the eighteenth centuries the lower, *wealthy* gentry and the growing commercial elites *were able to buy positions of political authority* either through the purchase of peerages from the Crown, or, especially later, through the purchase of 'rotten' parliamentary boroughs.
>
> In contemporary Thailand the business elite is largely Chinese, not Thai, and thus formal positions of authority are not open to it. Instead, members of the Chinese business community have established fairly stable relations with individual clique leaders in the Thai military in order to protect and advance their entrepreneurial

concerns. Relationships are of course, enormously rewarding for members of the Thai bureaucratic elite who oversee the licensing and taxing of enterprises, and many of the transactions that provide the cement for such relationships are quite illegal. Deprived of formal office holding, Chinese businessmen in Thailand have nevertheless managed well albeit through corruption, *to share fulsomely in the decisions which affect them.*[51]

Corruption, at this level, is thus a form of informal political influence on the part of those groups which hold considerable wealth, and yet for some reason do not, or do not yet, have legitimate access to the political arena. The reasons differ in the case of seventeenth-century England and contemporary Thailand (and other developing countries). Whereas the former was a case of an uneven, discontinuous evolutionary process, the latter is a case of a distorted evolutionary process. The Industrial Revolution in England had seen the rapid rise of new groups with new resources who naturally were trying to make themselves effective in the political sphere. At the same time, the political systems had not yet caught up with this new situation and hence did not provide legitimate and acceptable means for the articulation of the interests of those new groups. They had to await the ripening of responsible and sophisticated political parties and the graduation of parliamentary democracy.

In today's underdeveloped countries the source of the discrepancy between possession of wealth and access to political power is different. As we have observed before, the colonial regime and the internationalisation of capitalism interfered with the evolutionary development of indigenous social structures by artificially introducing economic elites (pariah entrepreneurs) often of different tribal origin, ethnicity or race, to act as middlemen or 'compradors'. In this way colonialism helped foster polycommunal societies where considerable wealth is concentrated in the hands of groups who are either marginal by permanent design and intent, as in the case of ethnically and racially distinct groups, or marginal by temporary misfortune, as in the case of minority tribes who are simply waiting for their turn (or return) to political power. To make matters worse, *foreign* commercial and industrial interests often dominate the underdeveloped economy, and these interests again seek illegal articulation through bribery.

The narrow entry into the political arena and the large economic interests of politically marginal groups combine to make for a staggering accumulation of wealth by those in power.

Some historians have viewed the growth of corruption in England in the seventeenth and eighteenth centuries as an alternative to violence.[52] Similarly, corruption in contemporary developing societies is popularly viewed as a form of conflict or tension management in an ill-adjusted social system. This point, for example, has been made by Huntington.[53] He argues that, whereas both violence and corruption are illegal ways of making demands upon the system, corruption is relatively peaceful, and where successful, functionally operates as a stabilising force. As long as wealthy elites can bribe the political elites into satisfying their demands, they will be less prone to use their money to finance violent attempts to overthrow the political regime. As such, corruption can be seen as the only means of integrating marginal groups into a disjointed social system.

Functionalist analyses, like the above, whilst opening a sympathetic perspective upon the causes of a serious malady in contemporary developing societies, must not, however, become a licence for appalling social injustices. We must not forget that (1) the majority of the people in underdeveloped countries do not have the financial resources to have their demands met by resort to bribery, and that (2) the corrupt back-scratching between political and economic elites increasingly blocks their access to the alternative, that is violence, as well. Many developing societies are ruled by the gun, and bribery pays for the bullets. Importation of arms into the developing countries increases at a rate of 10 per cent per annum, well over twice the average annual rate of growth of domestic output.[54] Massive financial injections from marginal economic elites, ruling parties, and sometimes the smallest of oligarchies, can help to keep private para-military bodyguards in power indefinitely.

Although, as we shall see in the next section, political instability – in the widest sense defined as the frequent change of governments – is popularly thought to be another symptom of underdevelopment, there are a large number of underdeveloped societies where the staying-power of the ruling elite makes a mockery of the contrast with the so-called political stability of developed societies, where the average life expectancy of democratic governments since the Second World War is barely a year or two. In a world where money is all that is needed to assure anybody access to superior arms technology with which to keep the exploited masses suppressed, corruption ceases to be functional in any but the most cynical meaning of the word.

(2) *Extortion at the law-making level: consolidation of power by political elites.* Let us now, then, turn to the politicians and the higher civil servants. I have classed both together in the law-making category, because on the one hand higher civil servants are to a certain extent involved in the formulation of laws and policies, but more especially because in developing societies, for reasons that we shall discuss below, there is little differentiation between political and administrative functions in the higher echelons of government. Indeed, the bureaucratic elite is the political elite.

The most frequent form of corruption that springs to mind when talking about politicians is that which takes place during elections. Elections are the *locus classicus* of bribery and corruption. In order to get into power, and to remain in power, politicians may well have to bribe whole electorates. This appears to have been the case in the modernising situation in all societies, not just contemporary developing ones. Records have been kept of the huge sums of money involved in elections in nineteenth-century Britain. These are disclosed in a detailed account of the history of British electoral corruption by Wraith and Simkins.[55] The cost, for example, of the 1880 election came to £3 million. The 1883 Corrupt and Illegal Practices Act which brought formidable penalties for such practices put an end to that, and the cost of the 1883 election was down to £780,000. However, anti-corruption bills had been passed through parliament before to no avail since, in fact, the beginning of the nineteenth century. It was only the growth of the electorate (Gladstone rather shrewdly increased the number of voters from three to five million thus making the costs of bribery quite prohibitive), its political maturity, and the emergence and institutionalisation of responsible and impeccable political parties which helped establish a formal code of electoral conduct, that put an end to the long history of corruption in Britain.

In contemporary developing societies Western political institutions such as universal suffrage, free electoral competitions, political parties, and 'parliamentary democracy' have been introduced from above, as a colonial farewell gift so to speak, and are not particularly in keeping with the political culture of the masses of the population. Being uneducated the mass of the electorate is unlikely to understand its political rights and duties in the new nation, or the difference between one electoral promise and another. But being poor they do understand the language of money.

The sums required for having to bribe one's way through the ballot-box, in turn, makes it imperative for would-be politicians to accept and extort bribes (especially from marginal economic elites) in exchange for political influence once they *are* in power. And so corruption turns full circle.

Elections alone, however, do not constitute the only reasons for politicians' need for cash. They do not occur sufficiently often to be given the blame.

Rather, the root of corruption for this second category lies in the 'prismatic' nature of the political order in contemporary developing societies, which is one *where personal cliques, not formal parties, compete with one another.* Following on from the theoretical spadework of Huntington and Fred Riggs, James Scott develops this theme elaborately in his small but succinctly written volume, *Comparative Political Corruption.*[56] Briefly, his argument is that in modernising societies, including Britain in the seventeenth and eighteenth centuries, *patron–client* relationships are the basic tissue of the political structure. In a period of transition when traditional status-based rights are being eroded and a new institutional order is slow in being firmly established, the individual's social environment lacks stability. Personal security is precarious, property insecure, and in the absence of stable expectations and impersonal assurances of strong institutions, authority remains personal and subject to discretion and caprice. Protection and advancement in this context depend on the establishment of personal ties of loyalty with those who are in a position to do one the most harm or good.[57] The weakness of formal standards of procedure (for example of selection and promotion on the basis of universalistic criteria of qualification) coupled with the oligarchic concentration of political power, makes personal security and career fortunes contingent upon personal alliances with highly placed individuals. Furthermore, I would add that the ubiquity of the development ideology has wheeled government and administration to the forefront as agents of development, thus ever widening the scope of the government's bureaucratic activities. Hence the government is easily the largest single employer in contemporary developing societies. No wonder that in contemporary developing societies the bureaucracy has become *the* political arena *par excellence.* Ambitious young officials, knowing that advancement hinges upon personal ties, seek to become protegés (clients) of a high-level patron (or 'godfather', as I have heard them called in West Africa since the release of the film of that name). In return for personal loyalty and service, the protége expects to be shown favours over non-clients and to share in the good fortune of his protection.[58]

For the patron, the absence of fully developed political institutions means that *his* struggle for power depends, in turn, on his ability to assemble as large a following of clients as possible. And this is where we, finally, arrive at the root of political corruption:

> Beyond the all-too-common desire to become rich and reward friends, it is this clique structure of elite based politics that promotes corruption and patterns it in certain ways. Within the ruling elite, cliques, as the units of political competition, are created, maintained, and expanded by manipulating the prerogatives of state officers within the bounds of legality; much of it, however, is distinctly corrupt. The distribution of high posts, financial opportunities and government-controlled privileges represents the major stakes of political competition and also provides the adhesive agent for competing cliques.[59]

From my own observations I found that when competition between cliques becomes fierce, the accusation of nepotism and favouritism, of the very kind described above, becomes a deadly weapon in the fight. Under these circumstances the patron may increasingly have to strengthen the ties with his personal followers by making cash payments rather than favours that are likely to catch the public eye. A financial gift is much less conspicuous than the allocation of a government flat, or the promotion to a senior position. Corruption itself thus becomes increasingly monetised, and the vast sums required for maintaining one's personal following largely come from the group mentioned in the first category, the marginal economic elites. These marginal elites, in contemporary developing societies, it must be remembered, are, by and large, a different social group from either patron or clients. For they are not within the political arena, but outside it.

(3) *Extortion at the law-implementation level: petty crimes of petty officials.* There is a world of difference between the high-level corrupt practices described thus far and those petty corrupt crimes which take place at the base of the bureaucracy, which involve the petty officials in their day-to-day dealings with the general public.

Several roots of corruption at the level of law implementation have been suggested. In one sense or other they all go back to Weber's classic distinction between patrimonial and rational bureaucracy. Briefly, the argument is that in the (preceding) traditional patrimonial forms of administration there was no distinction between public and private funds. Characteristic of such administrations was a system of decentralisation of authority, with a hierarchy of officials (or, as the case may be, feudal lords) each collecting a share of the produce of their domain, retaining part of this share and passing up the rest to the official or vassal next in the hierarchical order, and so on until the final tribute reached the crown or overlord. Riggs refers to this system as the 'trickle-up system'.[60] Since private and public funds are not separated, this system does not leave room for 'corruption' – it only calls for abuse. The modern rational bureaucracy, on the other hand, is characterised by strict separation of public and private funds. Thus there is no direct relation between the salary of the tax collectors and the revenue he collects. Public funds are collected, passed on to the administration *in toto*, and redistributed from the centre. A 'trickle-down' system rather than a 'trickle-up' system!

Two things occur in the transitional prismatic situation.

(*a*) The two procedures are temporarily confused (and misuse follows) if only because the uneducated illiterate traditional peasant does not know the difference, and thus makes a too-easy prey for extortion. It takes a long process of education before the uneducated peasant will learn to see the tax collector or the police officer, or any other petty official, as a functionary who is just as much subjected to the same 'universalist' rules and regulations as he makes out his client is. For a long while the illiterate masses will continue to regard the officials as representatives of public *authority* rather than as public *servants*.

McMullan takes this as the core of his theory of corruption, namely the difference between the literate public servant and the illiterate general public.

> The farmer is relatively a child. He is uncertain of the exact contents of the various laws that affect him, and uncertain how he stands in relation to them. He knows he should have a licence for his shot-gun, but cannot be sure that the one he has is still valid, or if the clerk who issued it cheated him with a worthless piece of paper. He knows he should have paid his taxes, but he has lost his receipt, and anyway there is something called a tax year, different from a calendar year which 'they' keep on changing, so perhaps he should have paid some more anyway. Even if he feels he has committed no crime he cannot defend himself against the policeman.[61]

Furthermore, in many parts of the world before the onset of the modern era, the customary exchange of gifts was a normal and integrated part of social behaviour, and with the change-over from a patrimonial or feudal system of administration to a rational, modern bureaucracy 'legitimate gifts' fade into 'disguised bribes'. That is 'the offering of gifts to (visiting inspectors), or even the tribute to the traditional superior, becomes an exaction by an official who has no shred of traditional entitlement to it! in other words, a racket. But most people, including the victims, are not aware that this is a racket.'[62]

(*b*) If the general public is an easy prey for extortion in the transitional situation, the petty official's need for cash is equally explicable in terms of the conditions of the prismatic situation. It is often argued that in developing societies the salaries of civil and public servants do not stretch enough to meet the demands of their kith and kin. This is typically a case of a coexistence of mutually incompatible social elements of which corruption is the sad but inevitable result. Temporarily there is a carry-over into the wage-earning economy of a vast network of family obligations appropriate to a simpler age. 'It has meant that a man living on an inelastic wage, in an inelastic house, in conditions – in other words – in which his only reasonable family obligation would be to his children and his parents – is still having to assume the obligations appropriate to the village.'[63] A classic novel that illustrates this point very clearly is Chinua Achebe's *No Longer at Ease*.[64]

(4) *Bribery at the law-implementation level: easing the pains of transition.* Finally, we turn to the bottom of the pyramid of corruption – the general public. What motivates the general public to bribe officials? The fact that their traditional attitudes and ignorance makes them easy game for the extortionate practices of petty officials does not mean that the public itself is not often equally eager to use positive inducements to persuade bureaucrats to dispense certain services and privileges. Here, too, we may trace the roots of corruption to the prismatic situation, in this case to the divergence between a modern government and a by and large traditional public, and to the conflict that arises between new laws and popular attitudes. In contemporary developing societies, new rules, policies and laws typical of a 'diffracted' and differentiated social system are superimposed upon customs, mores and practices typical of a 'fused' situation. And, obviously, the desire to circumvent or break the law is greatest when this law is contrary to accepted social behaviour and opinion, or when this law is simply not understood, or is irrelevant in the context of accepted social behaviour. For example, in countries where 'petty trading' has been an established social (as well as an economic) custom for centuries, the introduction of licences to regulate the trade, and to guide it into more productive channels, appears to the general public as a nonsensical or even offensive interference with social behaviour.

In the course of our own history, laws – as a rule – have come as a rubber-stamping of long-established practices. In contemporary developing societies, however, the government imposes laws deliberately to transform social behaviour, not to transfix it. Of necessity, it has to prescribe laws which deviate from accepted social behaviour, for no society can be changed without laws that go against the interests or accepted practices of some, or indeed most, people in it. And, therefore, new laws will set up the kind of conflict which give rise to corruption.

Corruption: Good or Bad?
Even when one's main objective is to *understand* typical features of developing countries by placing them in their situational context, one is still constrained to evaluate them, if not in moral terms, at least in terms of their instrumentality with respect to the development objectives.

The assertion that corruption is 'functional', in that it provides a means of negotiation of the position of certain politically disadvantaged yet economically strong communities in a polycommunal society, should be treated with scepticism. It is true that corruption can help maintain the social order by offering an alternative to violence. In that sense corruption is 'functional' to the stability of the social system. What is forgotten in such arguments is (1) that the social order is a disjointed, unintegrated one, and (2) that it is supposedly committed to *change*. Therefore, an evaluation of corruption should consider whether corrupt practices help or hinder the transitional process. And, from our discussion of corruption in high places, it seems to me that here, at any rate, corruption is 'dysfunctional' in that it acts as a stagnating force, keeping in power undeserving elites.

It is sometimes argued that in underdeveloped countries where domestic capital formation is low, corruption in high places may help the accumulation of capital in the hands of a few potential investors.[65] Such nice economic theorising, however, evaporates in the light of harsh practice. Political elites in underdeveloped countries do accumulate capital, but because of the fragility of their power base in an unintegrated and politically unstable society, because of the constant threat to their political survival, and the near-sure knowledge that the day of retribution will come when all their illegal fortunes will be taken away from them, they are rarely inclined to invest their money at home, other than in the unproductive expenditures to consolidate their power base. Thus substantial fortunes leave the underdeveloped countries to the safer havens of Swiss banks and international capital markets where they are of no further use to the national economy. This being the case little 'positive' can be said for corruption at the top of our pyramid.

The situation at the base appears quite different. No doubt to the 'diffracted' mind the petty corruption of public servants is irritating and inefficient. Let us not forget, though, that the sums involved here are quite harmless especially because they are spread over a wide base and do not concentrate to cement the power of a particular elite. Thus petty corruption presents no threat to the democratic process. And, by allowing the general public some scope for the negotiation of unpleasant laws, petty corruption may even be appreciated as 'functional' in mitigating the stresses of transition.

THE PRISMATIC POLITY: CHARISMATIC LEADERSHIP, MILITARY REGIMES, INSTABILITY AND STAGNATION

Following Parsons we have defined the polity as that subsystem of society which 'organises collective action for the attainment of collectively significant goals' (see p. 22). This very definition implies the existence of two conditions, both of which are typically absent in the prismatic situation, namely

(*a*) consensus as to what collectively significant goals are, and

(*b*) an effective organising unit capable of mobilising and enforcing collective action.

Whereas the first condition raises the problem of the formulation of collective goals as an orderly outcome of the political process, the second raises the problem of legitimation of authority which will have to ensure the effectiveness of government.

(*a*) The political process is that process whereby the various subunits (groups and/or individuals) of society articulate their interests in terms of demands on societal resources (for example incomes, roads, water-wells, educational facilities and jobs). These various resources are created, in turn, by collective action. Ideally, therefore, there must be some link between these multivarious demands and the formulation of collectively significant goals on the one hand, and the feedback from collective goals through collective action to satisfaction of the original multivarious demands on the other.

In small 'fused' societal systems, characterised as they are by little functional differentiation and by 'mechanical forms of solidarity', there are unlikely to be insuperable conflicts of interest between different sections of the community, so that it is hardly surprising that consensus may be relatively easily obtained by 'talking until they agree' (see p. 105) especially if the authority of the chief is such that people take his word for it when he sums up the mood of the meeting. And if some disaffected section of the community, say a particular descent group, still does *not* agree, then the fused character of societal functions makes it possible for it to break away and set up its own community elsewhere.

By contrast, in highly differentiated societies there exist functional subgroups with a high degree of autonomy towards each other and towards the government. Here, the formulation of collective goals is the outcome of the competition and negotiation between groups none of which are strong enough to obtain absolute power, and, moreover, all of which have a basic loyalty towards the larger unit that encompasses them all – the nation state. Furthermore, because of their functional interdependence, they have developed some measure of tolerance towards each other, for they all recognise that none of them is self-sufficient enough to break away. The constitution of the nation-state is the embodiment of this recognition. It lays down the formal rules and procedures under which the conflicting interests of the various groups may be negotiated and horse-traded: for example the parliamentary system, electoral rules, provision for autonomous political parties and so on, help to minimise the conflict between interest groups within the political arena. The political fight itself is formalised.

Two things are therefore necessary to bring the polity into line with a state of society which has undergone, or is undergoing, economic and social differentiation: first, the institutionalisation of the political process, and secondly, the existence of a basic loyalty towards the state. The problem in developing countries is that the polycommunal nature of the social order, and the 'clect' character of its social organisation hinders the development of each of these conditions.

Huntington, who has given a classic description of the concept of institutionalisation of the political process in modernising societies, defines it as that process whereby political organisations and procedures acquire value, stability and a certain measure of autonomy.[66] By value and stability he means that they come to *outgrow people and events.*

Outgrowing people means that the procedures followed within the organisation, and the roles assigned, survive the members of that organisation. Thus, rather than people making the rules, it is the rules that determine what kind of people shall be attracted to the political arena. The whole issue, here, is whether a party or a parliamentary system will survive its original founders. The eagerness with which many political parties in developing countries appoint their leader for life reflects the anxiety associated with the weak institutional character of the party. By contrast, in the Netherlands, one of the larger political parties, not so very long ago, called up applicants for membership in parliament by advertisement in the newspapers.

Outgrowing events means that the test of institutionalisation lies in the ability of a political party to survive beyond the initial specific goal for which it was founded. The organisational goals of a political party (as indeed for many other organisations) are usually highly specific in the incipient stage. In the case of political parties in developing countries that specific goal was 'national independence'. Once that goal had been achieved, however, most parties failed to adapt themselves to, and adopt, other goals. Fanon has bitterly accused the African political parties for their hopeless inability to cope with the post-independence situation.[67] They had never thought beyond that first goal, and were totally inept in defining other demands and mobilising their members for other interests.

Autonomy, finally, requires a degree of differentiation of the political sphere from other spheres. The political role of a candidate should be seen to be separate from his other roles. It is only when a political party, and its officials, are relatively autonomous in that sense that they can hope to amalgamate and articulate the interests of more than one social group. The ascribed entrance into 'clects', which – we argued – is the typical form of organisation of groups (including political parties) in developing countries, stands in the way of such autonomy. As a corollary, those political parties that do attempt to create a wider base frequently suffer from a splintering into political factions which once again reflect traditional social divisions (based on regional, linguistic and ethnic loyalties).

As regards the requisite of basic loyalty towards the state, everything which has been said before about the polycommunal character of developing countries and the consequent absence of an integrated 'societal community' is relevant in explaining why this basic loyalty is missing. *The* main political problem of developing countries is that of the transference of loyalties from smaller primordial groups to an abstract all-encompassing unit – the nation state. As long as the criteria for membership in this 'relevant' societal community are not recognised by the people, the people cannot be expected to participate full-heartedly in the democratic political process. As Jennings pointedly remarked: 'The people cannot decide until someone decides who are the people.'[68]

This is why *nation building* is seen to be so absolutely quintessential for developing countries. For the essence of nation building is the search for a collective identity which is coextensive with the territorial boundedness of the nation state; a collective identity, furthermore, that can become the basis for consensus, solidarity and the shared acceptance of a patterned normative order. This search frequently leads leaders of developing nations to dig up common histories with common heroes and a common cultural heritage. A first step thereby is often the change of the country's name, back to the traditional name by which the territory was known before European conquest. Because of the lack of coincidence between the territorial boundaries created during colonialism and those of the areas for which the original names are remembered, this sometimes leads to absurdities. In 1972, Sierra Leone was seriously considering a change of the country's name to 'Songhai', until someone with a knowledge of African history and geography wryly pointed out that the traditional state of Songhai had, in fact, been part of the territory of northern Nigeria. Because their colonial heritage is frequently the only *common* history of the peoples now sharing the same territorial country, the best approach to the creation of a common identity clearly seems to lie in a revival of the recollection of that heritage, however unpleasant.

(*b*) The need for a basic loyalty and commitment to the nation state is closely bound up with the second requisite of any modern polity, namely that of an effective and legitimate government. Obviously, the consensual outcome of the political process and the effectiveness of a legitimate government are related, for only a legitimate government can command the loyalties of the people, motivate them into making sacrifices, and enforce those laws necessary to transform the productive base of society.

A core problem faced by new nations is precisely what Lipset has called the *crisis of legitimacy*. Thus he writes that 'the old order has been abolished and with it the set of beliefs that justified its system of authority. . . . The new system is in the process of being formed and so the questions arise: To whom is loyalty owed? And why?'[69]

According to a standard classification made by Max Weber, which has yet to be invalidated, there are essentially three ways through which an authority system may gain legitimacy, that is may obtain the 'accepted title to rule'.

(1) Through *tradition*, through always having possessed it and through reinforcing the belief in its rightness by various symbolic acts. The terms of reference of this legitimation are therefore sacred and supernatural. A traditional ruler is one whose powers are believed to

have derived from the gods, from whom he or his forefathers or his lineage have descended.

(2) Through *rational-legal definition*, when those in authority are obeyed because of a general acceptance of the appropriateness of the system of rules under which they hold office. This type of legitimation, of course, begs the question of the institutionalisation of the political process.

(3) Through *charismatic appeal*, when authority rests upon faith in a leader 'who is believed to be endowed with great personal worth: this may come from God, as in the case of a religious prophet, or may simply arise from the display of extraordinary talents'.[70]

According to Fred Riggs, the self-appointed charismatic leader is the logical intermediary between the traditional type of authority belonging to the 'fused' society, and the rational-legal type of authority belonging to the diffracted stage: 'A charismatic leader manifests powers (and capabilities) which are formally mundane, but informally and latently supernatural. To the more diffracted conscious mind, he is a mere man to whom authority is delegated, but to the residual, fused unconscious, he is possessed by occult powers which compel obedience.'[71] Whilst the sacred leader is recruited ascriptively by hereditary succession and the secular leader by appointment or election, the charismatic leader of the intermediate stage is typically 'self-appointed'. Once having seized power 'by violence, by cunning, by intrigue, or even fortuitous chance, he has to comply with the formalities to make it legal'. Thus the 'tutor', as Riggs calls him, 'proclaims his own mission to prepare others for self-rule, yet acknowledges that, formally his authority rightfully comes from the sovereign people over whom he rules'.[72]

Lipset, too, sees charismatic authority as well-suited to the needs of newly developing nations. For charismatic authority requires neither time nor a rational set of rules and is, moreover, highly flexible. Such a leader, says Lipset, plays several roles.

(1) He is first of all the symbol of the new nation, its hero who embodies in his person *its* values and aspirations. The classic example here, which Lipset quotes, is that of George Washington, of whom it was said that: 'Washington's virtues are America's virtues rather than vice versa.' Chairman Mao, today, seems to perform exactly the same function in China.

(2) He (can) legitimise the state, the new secular government by endowing it with his 'gift of grace'.[73]

Charismatic justification for authority can thus be seen as a mechanism of transition, an interim measure, which persuades people to observe the requirements of the nation out of loyalty to the leader until they eventually learn to do so for its own sake. For this reason it is terribly important that charismatic tutors should sort out their succession well before their death or retirement.

Because charismatic leadership typically fails to separate the source from the execution of authority, it is, however, at the same time extremely unstable. Unlike in cases of traditional or rational-legal authority where those in power may be removed without fear of damaging the political basis upon which their power rests, the two are always intricately interwoven in charismatic authority. Part of the trauma of the Watergate scandal was precisely that the Americans feared that the impeachment of Nixon might do irreparable harm to 'The Office of the Presidency'. This anxiety in the last analysis was related to the fundamental uncertainty as to the type of legitimation of authority, namely whether it was charismatic or rational-legal.

Given the inability to separate the sources from the execution of authority, says Lipset, the charismatic leader must either place himself in a situation where he is not subject to criticism, that is a strong one, or he must transcend partisan conflict, and indeed encourage the institutionalisation of that conflict (government and loyal opposition).[74] Most leaders of developing countries seem to have opted for the first alternative, and even the most 'charismatic' amongst them, after a few years in office, find it increasingly necessary to suppress opposition by force. In the absence of a charismatic tutor, or in cases where the charismatic tutor fails during his lifetime to transfer the people's loyalty to 'a government of men under law' then the prismatic vacuum is likely to be filled by the military.[75] The army is a likely candidate for political power in newly developing nations for two main reasons: first, in the absence of any of the three forms of legitimacy, the army is the only group which can *enforce* legislation, and secondly, in stark contrast to the prismatic organisational shortcomings of the political parties in developing countries, the military is the only well-organised functional group: it is often a perfected bureaucracy; it has a good communications system, discipline and an *esprit de corps*, and, of course, it is well-trained to handle the tools of its trade. Today, one-fifth of the total world population is living under military rule, and with the exception of Spain, Portugal and Greece, they are all in developing countries.

POLITICAL INSTABILITY OR STAGNATION?

We shall wind up this chapter with a few paragraphs on 'political instability', which is so frequently alleged to be a typical problem of developing countries. The classic texts on this subject have been written by American political scientists amongst whom Huntington, Pye, Apter and Eisenstadt are the most important.[76] Their frame of analysis which led them to examine and explain political instability in developing countries closely resembles the kind of functional analysis of the prismatic situation presented in this chapter.[77]

What exactly is meant when speaking of political instability? Political instability, says Huntington, is marked by a rapid sequence of governments, of governments, moreover, that may not be different in orientation and formulation of societal goals, but which are simply different in that they are recruited from different elites. That difference is painfully brought out by the fact that, nearly always, *coups d'état* seem necessary to bring a new group to power rather than the constitutional procedures formally operative in these countries. Huntington, writing about the twenty years since the Second World

War, notes the increasing incidence of *coups d'état*, guerilla insurgencies and military uprisings in 'many modernizing countries of Asia, Africa and Latin America'.[78]

It has been the merit, I think, of political scientists like Huntington, Pye, Apter and Eisenstadt that they so clearly trace the roots of 'revolt' not in the situation of backwardness itself, but in the modernising or developing situation. As Huntington succinctly puts it, 'it is not the absence of modernity but the effort to achieve it which produces domestic violence and instability. If poor countries appear to be unstable, it is not because they are poor but because they are trying to become rich.'[79]

This modernising situation is precisely what Riggs has called the 'prismatic' situation. Modernisation, says Pye, is that process of social change in which tradition-bound villages or tribal-based societies are compelled to react to the pressures and demands of the modern, industrialised, urban-centred (indeed 'diffracted') world. Urbanisation, literacy, education and the mass media all expose the traditional man to new forms of life, new standards of enjoyment and new possibilities of satisfaction. These experiences break the cognitive and attitudinal barriers of the traditional culture and promote new levels of aspirations and wants.[80] This process of new-want formation has been called, since the pioneering work by Deutsch, *social mobilisation*.[81] However, the ability of the transitional society to satisfy these new aspirations grows much more slowly than the aspirations themselves. Consequently a gap develops between aspirations and expectations, between want-formation and want-satisfaction. And it is this gap, in turn, which generates *social frustrations*. Social frustration, again, leads to demands on the government and to the expansion of *political participation* to enforce those demands. Finally, what makes this political participation lead to political instability is the *underdevelopment of the political system*, as reflected in the lack of political institutionalisation and the absence of legitimation of authority, discussed previously.[82]

In the typically polycommunal society, the absence of a developed political system makes it impossible for some groups and new elites, and indeed the socially mobilised masses, to articulate their interests through 'normal political channels'. We mentioned in the discussion on 'corruption' that bribery may be one way of expressing those demands. But where bribery fails, political violence takes over.

This line of argument, to which American political scientists are especially prone, and which reflects their firm commitment to the functionalist-diffusionist approach which this chapter has tried to present, is – in my view – a reasonably correct interpretation of *some* political instability which has occurred in developing countries since independence. However, it is unsatisfactory on two accounts:

(1) It *never* mentions the instances of political instability which have been 'caused', not by internal prismatic or modernising conditions, but by 'destabilisation' policies of foreign metropolitan countries and their espionage agencies. In the post-war period, and after de-colonisation, the United States especially has acted as a world-wide 'policeman standing guard over vested interests' as the historian

Arnold Toynbee expressed it.[83] David Horowitz lists U.S. military actions (over the same period as that which is covered in Huntington's book on political instability) in Greece, Turkey, Iran, Guatemala, South Vietnam, South Korea, Lebanon, Laos, Cuba and the Dominican Republic; this is not to mention various African states where the C.I.A. has been known to have master-minded military coups (Ghana and the Congo).[84] All these countries, incidentally, appear on Huntington's list of unstable examples.

(2) More seriously, the problem with this functionalist approach lies with the identification of political instability both as a 'typical' empirical phenomenon in contemporary developing countries, and with the assessment of it as 'problematic'.

In the most recent *World Handbook of Social and Political Indicators*, comparative data on 'regular and irregular executive transfers' are reported for 136 countries of the world, covering a period of twenty years between 1948 and 1967. 'Regular executive transfers', the authors define as 'a change in the office of national executive from one leader or ruling group to another that is accomplished through conventional legal or customary procedures and unaccompanied by actual or directly threatened physical violence'. An 'irregular executive transfer', by contrast, is 'a change in the office of national executive from one leader or ruling group to another that is accomplished outside the conventional legal or customary procedures for transferring power in effect at the time of the event and accompanied by actual or directly threatened violence'.[85] Irregular executive transfers are what the layman commonly refers to as *coups* and they are interpreted by political scientists as manifestations of 'political instability'.

In the twenty years covered in the *Handbook*, 147 irregular executive transfers took place in a total of fifty-three countries only, the remaining eighty-three having experienced no such event. Out of the fifty-three countries only fourteen experienced four or more such *coups d'état* in that twenty-year period. Admittedly, with the exception of France (one recorded irregular transfer in 1958) the remaining fifty-two countries are commonly classified as 'developing'. But that still leaves fifty-six countries commonly classified as 'developing' with *no* irregular executive transfer in the entire twenty-year period.

On the other hand, the world has witnessed a great deal of regular executive transfers. A total of 1300, of which seventeen Western 'developed democracies' account for 188, and seven developed socialist states account for another thirty-seven, leaves 108 developing countries to share a total of 1065 between them. This gives them an average of about ten each, not much out of line with the average of just above eleven of the developed democracies. However, of the fifty-two 'developing' countries which had never experienced 'irregular executive transfers', twenty-two were also in the lowest quartile (that is 5 or less) of the distribution curve of regular executive transfers.[86]

What I have tried to demonstrate with this numerical jugglery is (1) that political instability as measured by 'irregular executive transfers', is not such a *commonly* occurring phenomenon of developing countries as political scientists, in their enthusiasm to explain it, believe it is, and (2) that in terms of executive transfers, that is change of government whether regular or irregular, quite a few 'developing' countries, in my opinion, could do a lot better. The *World Handbook*'s data are, moreover, understating my case, as they only cover the twenty years up to 1967. The developing countries of Africa had only obtained independence in the early 1960s, and their share of the world's *regular* executive transfers therefore mostly occurred under colonial auspices. Since independence, and quite contrary to widespread beliefs, emerging leaders in Africa have shown a suspect capacity for survival. Closely examining the thirty-six countries listed in the *Handbook for Africa South of the Sahara* (1974) we observe that between independence and 1970 *coups d'état* took place in eleven countries, often followed by counter coups, and from 1970 in seven only.[87] In January 1975, seventeen heads of state and/or government were in or well past their tenth year of effective power, and twelve of these had been in power since independence.* In Asia, the Middle East and in Latin America too, there are a surprising number of countries whose regimes have lasted much longer than the stereotype of political instability would have us believe.†

In conclusion, too much focus on political instability, coupled with a nicely constructed theoretical frame by which to explain it, diverts our attention away from what I believe is potentially a more serious problem of developing countries, namely *political stagnation*, by which I understand a relative lack of executive transfer. This phenomenon of political stagnation I suspect to be on the increase, due to the very success of corrupt back-scratching of political and economic elites – be they foreign or domestic – referred to before (see p. 132). Together

---

* These effective rulers are: Presidents Seretse Khama (Botswana, 1966); Micombero (Burundi, 1966); Ahidjo (Cameroon, 1960); Bokassa (Central African Republic, 1965); Tombalbaye (Chad, 1960); Jawara (Gambia, 1965); Sekou Toure (Guinea, 1958); Houphouet Boigny (Ivory Coast, 1960); Jomo Kenyatta (Kenya, 1963); Banda (Malawi, 1964); Moktar ould Daddah (Mauretania, 1960); Senghor (Senegal, 1960); Nyerere (Tanzania, 1962); Lamizana (Upper Volta, 1966); Mobutu (Zaire, 1965); Kaunda (Zambia, 1964); and Smith (Rhodesia, 1965).

In Liberia, moreover, President Tubman died peacefully, after thirty-eight years of ıeign, in 1971. In Niger, President Diori was ousted in a *coup d'état* only in 1974 after having ruled the country since its independence in 1960, whilst in Ethiopia Emperor Hailie Selassie took a long time in being finally overthrown in 1974 after an uninterrupted regime of fifteen years since 1960.

South Africa and the Portuguese territories are excluded from this analysis.

† For example in Asia the following autocratic rulers have enjoyed long survival rates: Mrs Bandaraneika (Sri Lanka, 1965); General Ne Win (Burma, 1962); Suharto (Indonesia, effectively since 1965); Marcos (Philippines, 1965); Lee Kuan Yew (Singapore, 1959); Chiang Kai Shek (Taiwan, 1948); Park (South Korea, 1953); S. I. Kim (North Korea, 1946–53).

In the Middle East and North Africa: Boumedienne (Algeria, 1964); Shah Mohammed Reza Pahlavi (Iran, 1941 – since 1955 more actively and autocratically); King Hussein (Jordan, 1952); Sheikh Sabah (Kuweit, 1965); King Hassan II (Morocco, 1961); King Feisal (Saudi Arabia, 1964–75); Bourguiba (Tunisia, 1956).

with the U.S. foreign policy of containment,[88] and its massive injections of military aid to strategic developing countries,* it allows repressive regimes and parasitic elites to stay in power long past their hour of popularity, and frequently in brutal violation of basic human rights. Political stagnation may also be detrimental to economic development as more and more national resources are channelled into maintaining the power base of an increasingly unpopular elite (for example expenditure on arms, military, security forces, police, para-militia, and so on). Furthermore, as leaders are well aware, economic development and modernisation (urbanisation, literacy and rising expectations) are potentially destabilising forces, so that in some instances economic development and modernisation may actually be discouraged by political elites.

---

* It is reported for example, that 80 per cent of U.S. exports of major weapons to the Third World have gone to countries bordering on the Soviet Union and China, and that until 1954 all U.S. arms to Third World countries were free of charge and since then 50 per cent have been free and a further 25 per cent subsidised or sold on easy terms under what Americans euphemistically call their 'aid' programme.[89]

Part Three
Development as Action

7
Alternative Development Strategies

In the first part of this book we discussed development as an autonomous process of growth and change of society, that is as a process of internal societal dynamics. Next we outlined the characteristics of underdevelopment of contemporary developing societies, as these had resulted from the historical interaction with more advanced societies. However, not only is underdevelopment, objectively viewed, the result of interaction with development, the subjective awareness and experience of underdevelopment also issues from the interaction with developed societies. It is from this interaction that the wish to develop – to catch up – is born.

Our next logical step, therefore, is to approach development from the point of view of 'action'. For, in the contemporary world, the word 'development' has come to represent a fashionable ideology, an ideology which reflects the active concern of governments for the growth and self-realisation of their societies. In this ideology, the word 'development' has become tantamount to planning, to the deliberate engineering of processes of internal societal dynamics, of growth and change. Note the distinction between government and society. To date 'development' is still very much an elitist ideology; a conscious effort and a pledge on the part of governments to achieve predetermined goals. Note secondly that the formulation of these goals is first of all *economic*, referring to an improvement in the material standards of living of the people. The reason for this is simple: just as the wish to develop, to catch up, originates in the interaction with the already developed societies, so too does the interaction with 'development' provide the main terms of reference for the various development models. As with the modernisation of Japan in the latter part of the nineteenth century, the active concern with development everywhere has always received its main impetus from the demonstration of the greater adaptive capacity of modern societies on the one hand, and from the fear of domination (or, as in most cases, from the actual experience of domination) by these advanced societies on the other.

The principal characteristic, and outwardly the most visible feature, of 'modern' societies is their economic and technological superiority. This superiority is demonstrated in the whole gamut of material provisions which they are able to offer their populations, including the poorer sections: adequate nutrition and medical aid resulting in higher life expectancies and lower infant mortality rates; good housing and the provision of public health and sanitation; education and cultural facilities; adequate military protection against external threats; plus levels of income which permit access to a wide range of consumer goods beyond the basic necessities. These are ever so many indicators of material standards of living which contrast sharply with the poverty, the squalor and disease, the malnutrition and under-nourishment, the high infant mortality rates and the low life expectancies, the overcrowding, illiteracy and the physical insecurity characteristic of the lives of the masses in the developing world. National leaders are acutely aware of this contrast, and of the way in which their underdevelopment makes them easy game for external domination and exploitation. It is therefore no surprise that in the definition of development goals the improvement of the material standards of living of the people is *invariably* formulated as the ultimate aim of development. At the same time it is correctly understood that in modern societies these high standards of living are closely associated with high productive levels both absolute and relative (that is both in terms of over-all output and in terms of output per unit of input). Whilst high standards of living result from high productivity, the obverse is also true: for example, adequate nutrition levels and good public health greatly improve the productive performance of labour. It is in this way that both an expansion of the productive capacity of the society and the improvement in material standards of living are the twin goals in any development model. Hence development as action stands principally for the inducement of economic development. But at the same time, national leaders, policy-makers, grass-root agents, and foreign 'advisers' have also become aware that, in order to achieve this goal, the entire socio-cultural framework in which the economy is embedded needs changing too. Thus 'development as action' has come to mean the deliberate engineering of both economic growth and social change.

One weakness of many contemporary development ideologies lies in the fact that, whereas the bearers of these ideologies have a pretty clear picture of what economic growth is, they do not have an equally clear notion of the desired direction of social change. In the West, as was observed in the first part of the book, the growth and the expansion of the economy came as the final fruit of a long process of structural differentiation and adaptive upgrading of the other institutional spheres of society in specific evolutionary order – political, religious, administrative and cultural, and juridical. The attendant integrative solutions which, through the evolutionary process, have given structural durability to the new forms of advanced societies, today appear as ever so many socio-political and cultural traits which contemporary developing elites are loath to accept. Whilst eager to imitate the outcome of the evolutionary process, that is high economic and technological adaptability, they hesitate to replicate the evolutionary steps leading to this outcome, first for fear of losing their identity, and secondly for fear of committing themselves to what, to them, often appear as inhuman values associated with the development of the First World. Development as 'action' puts development as 'process' on its head as it were. The stages of evolution are put in reverse order, and the conscious appraisal and rejection of these stages causes frictions and tensions at every turn. Contemporary developing elites are decidedly unenthusiastic about the way in which the adoption of the modern economic/technological complex is predictably going to 'steamroller' through the rest of the social order, inescapably turning their societies into replicas of the advanced societies.* Some leaders make conscious efforts to retain traditional institutions of family organisation and political participation in their economic modernisation, as Nyerere did in his creation of 'ujamaa' villages in Tanzania. Others take pains to defend the one-party state in response to stern reminders by modernisation theorists (and U.S. imperialists) that a multi-party democracy (rather) is a necessary corollary of economic development. Some, horrified by the competitive and exploitative individualism associated with the Western model, seek economic development through a total transformation of man into a creature which 'does not seek the path where advantage lies, but rather the path where duty lies', as in the ambitious socialism of Cuba.[1] Again others, more naively, think that they can preserve their cultural identity by simply not drinking alcohol (that most wicked of Western vices) as for example some of the Arab leaders who nowadays can afford to preach to us from the television screen.

Whilst generally in sympathy with those development ideologists who seek to achieve economic growth *and* retain (or, as in the case of newly emerging nations, establish) their own national and cultural identity, I must make my position clear at this point. We must recognise the validity of the functionalist modernisation theories which to my mind have conclusively shown that the organisation of economic life in any society is embedded in the wider context of the organisation of social life, and that, therefore, the transformation of the organisation of economic life has consequences for the entire social fabric. It is the avowed determination of the governments of developing countries to introduce economic development in their society *as a first priority*. Inasmuch as this very intention makes the organisation of economic life, and hence the to-be-introduced economic institutions, a *definitive* feature of the entire social order, one must consider old social, cultural and political configurations which do not functionally relate to these desired economic institutions as 'obstacles' to development; conversely one must view the new social, cultural and political institutions which are believed to 'fit' these desired economic institutions as both requisites and inevitable consequences of economic development. It is the *primacy* of the economic structure in the social order which marks the modern stage in social evolution, and it is this very primacy which limits the degrees of freedom in the organisation of society in the modern stage. This primacy of the economic structure, as we saw in Chapters 2 and 3, is expressed in the primacy of the principle of *economic rationality*, that is the desire to maximise benefits over costs, as the highest value governing social life and as the

* Cf. our discussion of modernisation theories in Chapter 3.

most general of human motives. Economic development stands or falls with the degree to which this principle is adhered to. Retention of traditional cultural values and social institutions always implies a 'cost' in terms of economic growth. Thus, for example, the desire to retain traditional institutions of extended family relations and loyalty, whilst they may indeed be combined with the pursuit of productive activities, can never allow for the most rational combination of factors of production in a society which itself has already transcended the boundaries of the kinship structure. It means tampering with the principle of economic rationality, for example in getting the right man for the right job, in allowing for the benefits of economies of scale, and so forth. Similarly, the desire to respect traditional communal ownership of land will always contravene the exigencies of economic rationality in that this may either block the transformation of land into larger and more efficient holdings, or prevent the land from being put to different uses despite, for example, the discovery of certain valuable minerals. Again, the desire to respect traditional religious values overlooks the fact that these religious traditions, in one way or another, imbue objects of reality (be they human, physical or technical) with a sacredness that diametrically opposes the principle of economic rationality, which rather requires a profane interpretation of objects, exclusively in terms of their instrumental value. And, finally, the consistent application of the principle of economic rationality also implies the treatment of human beings as mere instrumental objects. Within the structure of action dictated by economic rationality there is no room for human happiness, love and friendship, or the joy of work skilfully done. If technological advance and the benefits of economies of scale require that people relinquish their skills and crafts to a hand-operated machine, or to the debilitating rhythm of the assembly line, then that is just tough luck. It is this contradiction between the requirements of economic rationality on the one hand, and the desiderata of human values on the other that makes development ideologists long for an *alternative route*.

Although one is compelled to agree with the modernisation theorists regarding the over-all social structural and cultural changes which the primacy of the principle of economic rationality inevitably imposes, there is yet, I believe, one very important degree of freedom, one area of choice, which remains. This freedom of choice hinges upon the recognition of the ultimate purpose which is to be served by profit maximisation (that is by the application of the principle of economic rationality). And it is in this choice that the so-called 'capitalist' and the so-called 'socialist' development models take their respective points of departure. For, whereas in the capitalist model profit maximisation is an end in itself, in socialism profit maximisation is regarded as a means to other, *social*, ends, as the following quotations from a textbook on socialist economics written by a Polish economist, Oskar Lange, testify:

> Social rationality of economic activity demands that the aims of individual enterprises be subordinated to an end which embraces

> the whole of the social process of production and distribution; in other words, it requires the co-ordination of the activities of individual enterprises, the integration of their aims by a common end directing the economic activity of society. This co-ordination is called the planning of the social economy.
>
> The category profit is retained in socialist enterprise but ceases to be the ultimate end of its activity and becomes the means of subordination to the general social end of the plan. Profit serves as a stimulus to the completion of the planned targets and as a test of how far the economic principle is observed.
>
> From the capitalist mode of production the socialist mode of production, apart from productive forces, inherits only the methodology of the private rationality of capitalist enterprises, in particular calculation and book-keeping, together with the idea of the principle of economic rationality itself . . . in fact this principle becomes more important.[2]

Ironically, with this description of economic rationality (that is as profit maximisation by means of rational calculation and book-keeping) we are back with Weber's original definition of the capitalistic enterprise (see Chapter 2, p. 45) as, incidentally, Oskar Lange himself observes. Viewed from this definition, private ownership of the means of production is *not* an inherent feature of 'capitalistic enterprising', nor therefore of the application of the principle of economic rationality, and nor, therefore, is it an inevitable characteristic of economic development or the modern stage of social evolution. Because of their theoretical confusion of the *historic* development of the West with the *logic* of evolution, modernisation theorists have tended to assume that the mobility of the factors of production (land, labour and capital), which, to be sure, is a logical corollary of the application of the principle of economic rationality, depended upon the transferability of property rights (including rights over the means of production) to *private* owner agents, rather than to the state or any other organisation representing the collectivity. And, where – as in the 'capitalist model' – profit maximisation is accompanied by such private ownership of the means of production, then profit maximisation is bound to become an end in itself, namely for the private entrepreneur. Furthermore, in such a model of development, exploitation of man by man, as well as social inequality, is a necessary contextual development to which many will object on grounds of human values. For, as Marxists will point out, capitalist private property does not merely consist in the ownership of things but in a social relation between people:

> Property confers upon its owners, freedom from labor and the disposal over the labor of others, and this is the essence of all social domination whatever form it may assume. It follows that the protection of property (i.e. by the state) is fundamentally the assurance of social domination to owners over non-owners. And this, in turn, is precisely what is meant by class domination, which it is the primary function of the State to uphold.[3]

Surely, the reader will interject, if private ownership of the means of production is not a logical corollary of the application of the principle of economic rationality, and hence of economic development, and if, furthermore, state ownership of the means of production can avoid the exploitation, class domination and the inequality inherent in capitalist development, then the choice between capitalist and socialist models of development can hardly be an issue. Yet this very choice *is* the most hotly debated issue in the context of development today. To a great extent, of course, the reality and the intensity of this debate is but a reflection upon, and a rationalisation of, the reality of global politics, more especially of the imperialist rivalries of the superpowers each of whom try to extend their own spheres of influence across the globe and/or contain those of their opponents. Thus, for example, it is in the interest of U.S. capitalism to maintain structures of free trade and capital transfer with the developing parts of the world in order to ensure its access to raw materials and markets. And such structures of free trade and capital transfer can only be maintained (other than by the application of military force) if the elites of the countries at the other end of the structural link, so to speak, are equally committed to the ideology of capitalism and do not go around socialising the means of production (by seizing foreign assets, freezing the remittance of profits and ignoring patent rights). Equally, apart from any imperialist expansionist designs which they themselves may have, it is in the interest of the socialist superpowers (the Soviet Union and China) to contain the economic and thus the military strength of the U.S. monolith by persuading the developing countries of the advantages of socialism.

But even if the debate between capitalist and socialist models of development is but an ideological reflection of the struggle between superpowers for world hegemony, it is still relevant to examine their claims and their arguments (and their internal logic), for just as ideologies are rooted in the structure of reality so they may in turn mobilise people to act either to preserve or to change that self-same reality, and their power to mobilise people for action depends, in the last resort, on their persuasive strength.

CAPITALIST VERSUS SOCIALIST DEVELOPMENT STRATEGIES

Both capitalist and socialist models of development promise accelerated economic growth, but whereas the attraction of the socialist model lies in its claim to combine its strategy of economic development with the retention of basic human values of social justice and equality, the capitalist model appeals because it claims *faster* rates of economic growth now with the promise of more material benefits available for distribution later. Let us briefly examine the validity of these claims in the context of development today, which for all but a few countries (China, Cuba) *effectively* takes place within the network of the global capitalist system.

Both capitalist and socialist models agree with each other that economic development implies *industrialisation*, not merely in the wider sense of the term as we have defined it earlier (see Chapter 2, p. 48), but also in the narrow sense of a shift of productive activities from predominantly agricultural and raw-material extraction (that is primary activities) to predominantly industrial (secondary) activities. Given that all developing countries, socialist or capitalist, are committed to some form of *independent* development, there are some firm grounds for their identification of economic development with industrialisation.

(1) There is the observation that, when comparing the countries of the world today, we find that the higher the general level of economic development of a country (as measured by its *per capita* income) the higher the share of the manufacturing industry and the lower the share of agriculture, both in terms of output and in terms of employment.[4] The implication of this is that the disparity in the standards of living between industrialised and non-industrialised nations will, perforce, make for unequal trade, since the products of the latter always form but a cost component of the products of the former. Given the historically inherited *initial* disparity in standards of living, trade between rich and poor countries will at best simply perpetuate the existing inequality between them. Short of a global political and economic order with effective powers to redistribute income from rich to poor countries, international trade, therefore, cannot be relied upon to bring development to primary-producing countries. If these countries want to 'catch up' they will have to industrialise themselves.

(2) Improvement in agriculture itself depends for a good part on manufactured inputs, such as fertilisers, insecticides, farm machinery, and so on. In order to increase efficiency on the farm one must start a factory, as the Chinese have understood literally, when they built factories in their rural communes.

(3) Industrial (secondary) activities have a much larger capacity to absorb labour than agricultural (primary) activities. This, of course, alone would be sufficient reason for countries with ever-growing populations to industrialise.

(4) Industrialisation raises the over-all levels of skills and stimulates communication and transport; in short, it has the kind of external economies that are lacking in agricultural societies.

If both socialist and capitalist models agree on the paramount importance of industrialisation both as means and as goal of economic development, they certainly do not agree on the methods for achieving it. In the following examination of the two models we shall use the analytical scheme which breaks the organisation of economic life down into three interrelated processes:

(*a*) the mobilisation of resources;
(*b*) the production of goods and services; and
(*c*) the distribution of goods and services for consumption.

Economic development implies a total transformation of each of these three processes simultaneously and in correspondence with each other. Because socialist and capitalist approaches have radically different points of departure, we may trace their different strategies through each of these three processes.

(*a*) THE MOBILISATION OF RESOURCES FOR DEVELOPMENT: CAPITAL VERSUS LABOUR AS DOMINANT RESOURCE

The rate and the size of economic development in any country at any given time depends on both the size and the mode of utilisation of its economic surplus. This economic surplus may be viewed as a residual factor: that which remains from total output after necessary consumption has been subtracted. This economic surplus is therefore investable surplus.[5]

It is easy to see that without investment in buildings, roads, bridges, railways, power stations, machine tools and blast furnaces, industrialisation and high levels of total and *per capita* income cannot be obtained. But the ability to save and invest is itself a function of the level of income. At low levels of *per capita* income, people are too poor to save and invest, and the consequent low rate of investment will result in a low rate of growth of total income. This observation is better known under the label of 'low-level equilibrium trap'.[6] It is in the context of this argument that we should place the strategy of resource mobilisation of the capitalist model. For the capitalist model sees the increase in size of the economic surplus (that is of the investable surplus) as a function of the appropriation of the previous economic surplus (that is of the previous year) by a relatively small class of people called capitalist entrepreneurs. Appropriation, it is thought, not only makes accumulation of the surplus easier, but also facilitates the transformation of the surplus into productive capacity, since the relatively small class of rich people will have far more income than they can possibly want to spend on consumption. Transformation of the economic surplus into *productive* capacity will, in turn, increase total output and hence the size of the economic surplus available for investment in the next economic period, and so on and so forth. Furthermore, it is argued, free competition and the regulatory operation of the price mechanism will see to it that profits will characteristically go to the most efficient and productive of the entrepreneurs, and in this manner will stimulate the growth of the economic surplus still further. As Keynes observed: 'Capitalism is the extraordinary belief that the nastiest of men for the nastiest of motives will somehow work for the benefit of us all.'[7]

One may well wish to question the desirability of such a model on social grounds, for inequality and class formation are its *essential* features, and few would want to repeat the human toll of the early phases of industrialisation in Europe where capitalist development, however successful in the end, was accompanied by an absolute reduction in the standards of living of the masses and by an unspeakable assault on human dignity.

But quite apart from such human considerations, there are also a number of things wrong with the economics of this model in the context of contemporary developing societies. The model assumes certain *a priori* conditions which, because of the interactional context of global capitalism within which economic development is to take place, happen to be entirely absent in contemporary developing societies.

These *a priori* conditions are (1) the presence of an entrepreneurial class with a passion for savings and productive investments, and (2) optimal conditions for competition.[8]

In Part Two of this book we learned how contemporary underdevelopment is not a state of backwardness prior to capitalist development, but is instead a particular offshoot of capitalist development. Underdeveloped countries are structurally dependent on the capitalist developed countries, and this, by definition, rules out their own independent capitalist development. For the interests of the wealthy native elites are intrinsically wound up with the capitalist interest of the metropoles, which makes it unlikely that these elites will ever act other than as a comprador class. For example,

(i) they are unlikely to invest their surplus in productive activities which compete with those dominated by foreign capital; rather their money is tied up with that foreign capital;

(ii) they emulate the life-style of the wealthy in more advanced countries, and this – because of the disparity in productive and consumption levels between advanced and underdeveloped countries – means that, contrary to what the model says, the native rich have an almost infinite capacity to consume. This, of course, reduces their savings potential;

(iii) under the unfavourable political and economic climate at home they will tend to want to invest their money either in quick profit-yielding activities such as the construction of residential houses, or in mercantile activities on the fringes of foreign capitalist interests, neither of which contribute to the expansion of the productive capacity of the local economy;

(iv) finally, worst of all, the possibilities of higher returns on investments in the advanced countries coupled with the reduced risks involved in such investments, siphons their investable surplus off to more promising havens abroad.

The second condition which is an *a priori* assumption of the capitalist model, namely that of perfect competition, requires the absence of monopolistic or oligarchic enterprises. These conditions indeed existed during the early phases of capitalism in the West, but this is not the situation anywhere in the capitalist world today. The presence of giant multinational corporations (some 200 of whom already control over one-half of the world's total industrial output) inside the local economies of developing countries makes it economically irrational for native elites to invest in independent productive projects. Therefore, even if the local bourgeoisie were totally dedicated to the independent industrialisation of their country, it would still not make sense for them to invest in it.

To accommodate this problem, capitalist models of development have suggested a variant of the model, namely *industrialisation via import substitution*. This involves the home replacement of an existing market for final consumer goods. Domestic infant industries are encouraged by protective-tariff and tax-concession policies. This approach has, despite initial apparent successes, generally failed; this is because (1) protection raises domestic costs thus discouraging these industries from building up a competitive position in the unprotected export market; (2) because underdeveloped countries typically lack the technological equipment and know-how needed to set up consumer industries, therefore protection of infant industries had to be accompanied by still more liberal importation of producer goods;[9] and (3) because industrialisation via import substitution has encouraged the very spread of multinationals, since – in order to protect future markets – multinationals would 'jump' the tariff barriers and set up local subsidiaries even if this meant operating locally at a loss over a sustained period.

It is the *economic* failure of the capitalist model in developing countries which has convinced socialist strategists that independent economic development requires a radical break with the world capitalist system.

The *socialist* strategy, therefore, mobilises the existing economic surplus first of all by *expropriation* of wealth from foreign and domestic capitalists and landowners, and the consequent elimination of the drain on current income resulting from excess consumption and capital removals abroad. This leads to an instantaneous increase in investable economic surplus, but, perforce, this can only be a one-off increment. After that they are on their own, so to speak, and since the socialist model does not believe in inequality and class formation, the economic surplus has to be mobilised through *full utilisation of man-power* coupled with a *ruthless constraint on consumption.*

Full utilisation of man-power involves full employment, even at marginal returns, the provision of a taxable income for all the population and enforced savings by taxation. This strategy makes possible the increase of productivity per man-hour by means which do not require heavy capital investment, mainly because of the use of disguised unemployment that exists in the rural areas, and by transferring the workers/peasants to industrial and infrastructural projects that require much labour and little capital (irrigation dams, canals, roads, harbour repairs, and even steel works, as in China). As the Chinese leadership is fond of saying, 'The people are our capital.' As a necessary corollary of full employment and incomes, and the enforcement of savings by taxation, food prices must be kept low, so as to avoid inflation which would occur as a result of a sudden increase in effective demand, as well as to encourage savings. This can only be done through the *collectivisation of agriculture.* Thus, in this model, at least in the initial stages, industrialisation is mainly financed by agriculture. Central planning is a technical device which replaces the operation of the price mechanism in the capitalist market model. The country's economic surplus is transformed into productive capacity by techniques of differential costing and pricing of state enterprises, and especially through the manipulation of the terms of trade between agriculture and industry. Provided full statistical data are available, central planning can prove a better regulator of economic life than the market under capitalism: while under capitalism maximisation of profits does not necessarily imply the use of the best methods of production, under socialist planning one does not perhaps get the best profit but one can afford the best method of production. And with this we arrive at the discussion of the second process of economic life.

(*b*) THE PRODUCTION OF GOODS AND SERVICES: CAPITAL-INTENSIVE VERSUS LABOUR-INTENSIVE TECHNIQUES

In the capitalist model of development, the link between capital formation and economic growth hinges upon the conception of 'stable over-all capital–output ratio'. The capital–output ratio is the amount by which a given increase in investment increases the volume of output. Capital–output ratios, of course, vary between projects and between sectors of the economy. Thus the capital–output ratio for a textile plant is lower than that for a steel plant. But by averaging out the different capital requirements between the different sectors of the economy, one can – for purposes of convenience if not accuracy – arrive at the concept of an over-all or 'aggregate' capital–output ratio. The rate of growth of national income is then equated with the savings ratio divided by the over-all capital–output ratio. Since the history of early West European industrialisation suggested a *stable* capital–output ratio *over time*, the logic of the capitalist model was simply that the more capital a nation saves and invests, the greater its annual output will be.[10] No wonder therefore that capital formation, as well as the formation of a capitalist class, occupies such a central position in the capitalist development model.

However, an increase in output need not necessarily be a function of the increase in capital inputs, but may also be a function of *a change in the combination of the factors of production*, as Schumpeter, a capitalist economist, has already pointed out.[11] Technological innovations may in fact reduce the amount of capital input needed for a given output stream, or may increase the output relative to the same input. Conversely, technological innovations may reduce the labour required for a particular output, or indeed may reduce the amount of raw-material inputs required for a given output stream. For example, certain forms of technical reorganisation of the work process, like the introduction of multiple work-shifts in factories, greatly reduce the capital–output ratio, whereas the introduction of assembly lines, trucks, bulldozers and telephones reduces the amount of labour required. Thus, labour-saving techniques are often, perforce, capital-intensive, whilst capital-saving techniques are often labour-intensive. The earlier-cited capitalist models which had emphasised the importance of capital accumulation in development were correct in suggesting that in the historical development of the West the aggregate capital–output ratios had remained fairly stable. This was not to deny technical progress and innovation. It just so happened that in the development of Western Europe and North America the ruthlessly regulatory operation of the free-market system contributed to the complementarity of the factors of production. As a consequence, capital-saving and labour-saving innovations had taken turns and so offset one another. The upshot of this was indeed a fairly constant over-all capital–output ratio. For the operation of the free-market system would manipulate the supply of labour and capital, and this in turn would influence the type of technical innovation to be made. When capital was dear relative to the price of labour, entrepreneurs would seek to introduce technical devices designed to absorb a lot of labour and save capital. But this would then, by itself, gradually lead to a greater demand and higher prices for the factor labour until the scales tipped once more in favour of capital. The introduction of capital-intensive production techniques would follow and the demand for labour would diminish until its price had once again reached a level where it could successfully compete with capital, and so on and so forth.

In contemporary developing societies there is, however, an important structural feature not present during the capitalist development of the First World, and that is the *permanent oversupply* of labour, due to the population explosion which, as we have observed before, has itself been the result of contact with the advanced West. The existence of permanent surplus labour limits the applicability of the above-mentioned theory, which assumes complementarity of the factors of production. A permanent labour surplus obviously calls for a *bias* in the type of technological innovation that affects the production functions, that is a bias in favour of labour-intensive and capital-saving techniques. To date this problem of 'alternative', or 'intermediate', or 'appropriate' technologies, as it is variously referred to, for the developing countries is one of the most widely discussed topics in development economics. China is often quoted as a good example of a developing country that has systematically tried to 'slant' its production process in the direction dictated by its relative factor supplies. For example, in China road-building is done by sufficient people with shovels so as to replace bulldozers, and insecticide sprays are replaced by hand-weeding. However, as Frances Stewart has very rightly observed, the choice of technique depends on the choice of product.[12] Developing societies which stand in an open-trade and capital-transfer relationship with the advanced Western countries cannot control the patterns of consumption, and therefore they cannot control the type of producer goods used in their countries. Labour-intensive methods in, for example, textile manufacture or in brick-making, Stewart argues, are excluded when modern-style textiles or bricks of standard and homogeneous strength and appearance are produced. The dependence on trade with, and aid and investments from, the advanced world thus leads to what Stewart calls 'overkill' needs of the underdeveloped countries. In socialist developing countries, by contrast, where there is domestic political control over patterns of consumption it is possible to avoid inappropriate products and therefore also inappropriate technologies. But, we must hasten to add, such domestic political control over consumption, and the ability to keep patterns of consumption in line with local productive-capacity, requires a degree of isolationism and self-sufficiency which only the largest of developing countries can hope to achieve. For only those countries which have both a large potential domestic market as well as sufficient natural resources can hope to develop a minimum degree of contact with the already existing patterns of world trade and production.

A development policy that substitutes capital by labour would seem to kill two birds with one stone, as capital is short and a more labour-intensive technology will provide full employment and also, therefore, social justice and greater economic equality. Labour-intensive techniques are hard to come by today though, due to the relative stage of technological advance already achieved in the West. But the real dilemma stems not from this, but from the fact that capital-intensive methods are often more advantageous from an economic-growth point of view, as for a given amount of capital investment they yield a higher net output or surplus which can then be made available for further investment. So, although they yield a lower volume of present employment and consumption, they promise a higher rate of growth, and therefore a higher potential level of employment *and* consumption in the future.[18] This is one of the 'attractions' of the capitalist model.

With the presentation of this dilemma between *growth and equity*, we have now come to the third and last economic process.

(*c*) THE DISTRIBUTION OF INCOME: GROWTH VERSUS EQUITY

The distribution of the national product involves the distribution of income to the three factors of production – land, labour and capital. In this manner the third process of economic life links up with the first: in any economy the distribution of income becomes the engine for mobilising the nation's resources for production in the next economic period. In a developing economy the core problem of planning becomes articulated by two main considerations:

(*a*) on the one hand income needs to be distributed such that this distribution will ensure the over-all most productive performance of the factors of production in the next economic period;

(*b*) on the other hand distribution of income needs to be such that it encourages the most effective demand for the produced goods. When defined in this manner, the whole conflict between 'distribution for growth' and 'distribution for equity' – at least for the larger of the contemporary developing economies – will be seen to disappear.

At the time of Europe's early economic development there was indeed a conflict between 'growth' and 'equity'. The reason for this lay in the circumstances which were apparent in the early days of European capitalist development, namely that *foreign* markets existed to create the demand necessary for the encouragement of production at home. Hence the classical capitalist model of development took no account of the constraint now bearing upon developing countries of having to create an internal domestic market. In the absence of such constraint, and under the assumptions of the classical capitalist model, distribution for growth and distribution for equity may appear to be at odds with one another. Because of the assumed relationship between concentration of capital, capital accumulation and economic growth, in the capitalist model the owners of capital are much favoured in income distribution over and above the owners of labour and the producers of primary inputs (agricultural workers). Early capitalist developments therefore kept wages low, as well as the prices of agricultural products, so as to be able to suppress the wages of workers in industry still further. This was possible because these same wages and rural incomes were *not* needed to create an effective market for the finished products.

The situation is very different in the developing countries today, since, as late developers, they encounter a world market already conquered by the early developers. As a consequence, for their own economic development they have to create internal markets to pull the economy out of its traditional stagnation. An equitable distribution of income is a means towards creating such a market, and this is the economic advantage of the socialist development model which encourages the creation of such a domestic market because of its full-employment policy.

But even for the *smaller* contemporary developing countries, that is those who for reasons of size lack a potential future domestic market, the conflict between distribution for growth and distribution for equity appears to be a spurious one upon closer examination. Social objectives aside, even from a narrow economic point of view, growth and equity need no longer be seen as contradictory goals, given the peculiar constraints operative in contemporary developing countries.

The original argument which placed growth and distribution at loggerheads with one another was based on the simple assumption that growth of income was a function of the aggregate marginal propensity to save which in turn was seen as the sum of the class marginal propensities to save weighted by the class income shares. Since empirical observation seemed to confirm that the high-income groups are indeed the high marginal saver groups, it seemed obviously true that redistribution of income from savings and high income to low-income groups would result in a diminished level or rate of growth of output and total income. Recent economic development theories, however, point out that what developing societies need is not just 'domestic savings' but 'foreign-exchange savings'. For developing countries typically rely on imported capital goods and intermediate inputs since these can generally be produced more cheaply in the more advanced countries. They need these imported goods for their own industrialisation and general development efforts. Therefore, foreign exchange takes on the role of a *separate scarce resource*, which in the short run is not perfectly substitutable for domestic savings. On the contrary, the very group which has the highest propensity to save also happens to be the group which 'drains' foreign exchange by its excessive demands for foreign consumer goods. Economic-development models which capture this aspect of growth of income of developing countries have been termed 'two-gap' models.[14] And it has been suggested that an income redistribution policy through taxation, in conjunction with a government policy orientated towards 'demand management' may yield, in fact, even higher rates of economic growth than policies that neglect income redistribution and demand management.[15]

In conclusion it would appear that both on social and on purely economic grounds socialist approaches to development have the edge over capitalist approaches in more than one way. However, to replace the invisible hand of the capitalist market with the visible directives of the socialist planners presents tremendous challenges to human moral behaviour. The capitalist model has at least one big advantage, namely that it tries to make the best of human nature at its worst. Man's competitive greed is used to work for society's over-all adaptive advantage. The nastiest of men, for the nastiest of motives, were after all delivering the goods. The socialist model, by contrast, requires the presence of saints. It requires a total dedication, and a total commitment on the part of the elite to 'serve the needs of the people', as the Chinese saying goes. The temptations for the power holders, that is the party, which commands all the nation's resources in the name of the people, is corrupting to the extreme. Rene Dumont's book *Socialisms and Development*[16] exposes the proto-socialisms of the national bureaucracies in most of the contemporary socialist states. Since the required level of morality and integrity does not come overnight, checks and balances must be brought in at the structural level, whilst education must cultivate the necessary ideological commitment. For example, great care is taken in China to ensure some structural grass-root control over the political process as well as over the behaviour of the party as a class. First, grass-root approval is an important criterion in the recruitment and admission of candidates into the party; secondly, the masses are represented as full members *alongside* party members in the revolutionary committees which run every social and economic organisation in the land, be they schools, factories, hospitals or communes; and thirdly, most importantly, a revolutionary *élan* is permanently maintained by the staging of cultural revolutions at regular intervals which are designed both to bring ideological commitments into line with economic realities and to prevent the fixation of class interests on the part of emerging elites (that is the party, technocrats, intellectuals, and similar groups). As for education in China, this begins in nurseries and ends in the grave. It is consistently and feverishly directed towards the creation of a 'new man' who will genuinely *want* to serve the needs of the people and who will be loath to seek his own self-interest. Nor is this education confined to formal schooling. Massive educational efforts are daily pumped into the factory, the commune, the theatre and the playgrounds as well as the schools, and in this way China just might succeed in creating a new breed of man – 'saints'.

8

Development Models: Ideology or Utopia?

In the previous chapter we reviewed the two dominant development models, capitalist and socialist, in their main outline.

I shall now make the perhaps surprising and unusual point that both these development models are *ideological* rather than *utopian* in character, for the substance of these two models derives from – however contrasting – interpretations of past reality rather than from an interpretation of future reality.

The distinction between ideology and utopia may come as a surprise in the context of a study of 'development'. After all, are the 'ideologies' of development not precisely about future society? So they are. Would therefore the term 'development ideology' not have the same meaning as 'development utopia', provided of course that we agree to relieve the word 'utopia' of its inappropriate connotation of unrealisability? Not quite! In conventional usage, both within the social sciences and in colloquial language, the term 'ideology' has become so over-used that it has lost the original terms of reference given to it by the inventors and the early users of the concept. The word 'ideology' now stands for just about *any* integrated set of ideas, beliefs and values which may propel people into action and justify their actions. Where these beliefs, ideas and values, in short, this consciousness, spring from, and what their relation to actual social reality is, no longer features in the definition of the concept of ideology.

Yet at one time in the history of the social sciences, the origin of 'consciousness' and its relation to actual social reality were seen as essential terms of reference for the definition of the concept of 'ideology'. And at least for one influential and authoritative writer on the subject, namely Karl Mannheim, these same terms of reference provided a clear demarcation line distinguishing between those forms of consciousness which were ideological and those which were utopian. It is hardly the place here to present a systematic survey of the concept of 'ideology' and the many meanings it has been given. On the other hand, a proper handling of the concept is pertinent to the study of 'development models', for it appears to me as the only theoretical construct which can assist us in the evaluation of both the relevance and the realisability of the development models that we have been discussing. But that means that we must, however briefly, get the terms of reference for the concept sorted out.

Fundamental to the concept from the start, has been the notion that consciousness (that is the content of knowledge about reality and the substantive ideas and values derived from this knowledge) is socially determined and rooted in the social structure of society. To the earlier 'ideologists', men such as Destutt de Tracy, Diderot, Condorcet and other members of the French Enlightenment, the contemporary structure of society appeared as a false and inadequate social order, and inasmuch as they considered the consciousness of the time to be a reflection of this false social order, they endowed this consciousness with the description 'false'. Ideology, therefore, was in the first instance 'false consciousness', and the critique of this false consciousness, perforce, turned into a critique of 'false' society.

With Karl Marx, who intellectually descends from these ideologists, a new element, a new dimension of meaning, enters the concept of ideology. Consciousness is false and therefore ideological, not merely because it reflects a false social order, although Marx concedes that it does, but *primarily* because it fails to recognise its own social roots. And, given that in any historical period the propagators of the dominant ideology are those who benefit from a continuation of the status quo, that is the upper classes, this false consciousness becomes even more invidious: it is not just a reflection, but a cover-up, a camouflage which masks the real social conditions (albeit that these 'real' social conditions are as yet 'false' from the perspective of true 'ideal' reality, namely the future classless society), and thus hindering the historical progression towards that true reality.

Throughout the history of Western philosophy – and the history of the concept of 'ideology' is no exception – the ultimate terms of reference within which philosophy has unfolded have been the same as those which have embraced the whole of Western civilisation and which we have discussed so extensively in Chapter 2. The observant reader may have picked them out already: it is the notion of *duality* – of duality between *this* world and the *other* meta-empirical world; between this historical 'real' reality and yonder absolute 'true reality'; between real forms and true substance; between that which is real in the *praxis* of everyday life and that which is true in scientific theory where we have stripped real phenomena of their apparent concrete characteristics to arrive at their abstract but 'true' qualities. This duality and the intellectual tension it creates has been the calling tune of Western civilisation: can that which is real ever be true? Will that which is true ever become real? Where the philosopher searches and suffers, the layman, in exasperation, exclaims 'But this is *really true*', a colloquial expression which to my knowledge is common only to Western languages.

One solution to this agonising problem of 'duality', and one which Karl Mannheim ingeniously exploits to the full in defining his concepts of 'ideology' and 'utopia', is the good old compromise. That which is real is part of the 'truth'. Historical realities are but manifestations in a dialectic unfolding of 'true' reality, for the structures in one epoch automatically call forth the structures of the following epoch. Marx, of course, had adopted a similar compromise, but whereas for Marx 'truth' awaited at the end of the sequel of historical unfoldings, for

Mannheim 'truth' was more evenly distributed over all its historical manifestations. And if only, Mannheim wished, one could transcend to some *ekstatisch Ausserhalb*, to some meta-historical point, out of space and out of time, one might pick up the 'truth' that runs like a red thread through history, connecting all historical manifestations and disclosing the 'coherent symptomatic' unity and sense of history.[1]

It is from this ontological compromise, that is contemporary reality as part of a dialectically unfolding history, that Mannheim derives the epistemological terms of reference for the definition of his concepts of 'ideology' and 'utopia'.

Mannheim accepts whatever his predecessors had said about ideology being false consciousness, but he adds the distinction that, whereas in the case of *ideology* the falseness of consciousness lies in its interpretation of actual reality in old and obsolete terms (terms that 'lag behind the times', as it were), in the case of *utopian* consciousness the interpretation of actual reality derives from interpretations or prognoses about future reality (running ahead of times, as it were). Both ideology and utopia are false consciousness when it comes to assessing their adequacy as descriptions of actual reality, but although in that sense utopia is just as 'false' as ideology, it has the possibility of being realisable, a possibility which ideology, perforce, never has.[2] The proviso is, of course, that utopian consciousness has the good sense and intelligence to pick out the red thread which runs through history and which connects up the sequel of historical periods. It has to recognise 'the makings' of the next historical period which – *in statu nascendi* – are already discernible in *this* historical period. If utopian consciousness fails to do this, it automatically becomes ideological once more. Whereas we must identify as ideological all beliefs, knowledge and values which can never be realised, it is the characteristic of utopian consciousness, precisely, that is realisable. And because it is realisable, it may have an *umwalzende funktion*, the power to effectuate transformations of the existing social order.[3] But its realisability, and therefore its powers of effectuating these transformations, depends on the degree to which it stays in tune with the dialectically unfolding historical process. How utopian consciousness can perform this monumental task Mannheim never makes very clear, at least not in his earlier work *Ideology and Utopia.* Referring to earlier definitions of ideology as the social determination of human thought, he suggests that as a first condition utopian thinking must be carried out by people who have no social (class) roots in actual historical reality. Mannheim naively suggests that a *freischwebende intelligenz* of scientists and academicians would be a likely candidate for the job because, as a functional group, they appear to have the widest recruitment base in society, and therefore are the least suspect from a point of view of class interests.

PRINCIPIA MEDIA

In a later work, however, Mannheim develops a procedure to be used by the *freischwebende intelligenz* – the scientists – whom he now has promoted to 'planners'. With this procedure they can keep their finger on the pulse of the historical heartbeat so to speak, so that they can predict the makings of the next historical period in truly 'utopian' fashion. This procedure is the identification of *principia media. Principia media* are the structures by which the general laws of events are mediated and expressed in the individual destiny of each historical period. They are 'universal forces in a concrete setting as they become integrated out of the various factors at work in a given place and a given time'.[4]

It is the task of the planners to separate the sheep from the goats, and to distinguish critically between the general laws of human society and the cloak of the *principia media* in which these general laws become *cemented* in any one actual historical period; at the same time the planners must predict the *principia media* of the next historical period. As an example of such obsolete *principia media*, Mannheim discusses a thesis developed by Max Weber with which the reader must be thoroughly familiar by now; the thesis, namely, that a formal legal system based upon rational and calculable principles, as well as fundamental equality before the law, is a necessary structural counterpart of capitalism. Proper social analysis of the kind propagated by Mannheim reveals that this general principle is not general at all but only a *principium medium*, limited to the liberal competitive phase of capitalism. For example, he writes that

> at an earlier period of capitalist development the contracting parties appear before the law with approximately equal strength at their disposal, but in a later stage of monopoly capitalism, the partners are of unequal political and economic power and we find an increasing element of judicial irrationality in the shape of legal formulae which leave the decision of the case to the discretion of the judge, dispensing with the old principles of formal law. Such clauses as the consideration of 'public policy', 'good faith', or the interests of the 'concern itself' give the judge a chance to disregard the formal and egalitarian application of the law and to open the door to the influence of the real holders of power in society.[5]

As Anatole France one remarked with witty sarcasm, 'The Law in its majestic equality forbids the rich as well as the poor to sleep under bridges, to beg in the streets and to steal bread.'[6]

PARSONS, MARX, MANNHEIM: TOWARDS UTOPIA

Equipped with Mannheim's distinction between 'ideology' and 'utopia', let us now identify the ideological nature of existing development models, both capitalist and socialist, and prepare the ground for a utopian development model. To this purpose, and in balance with the generally eclectic format of this book, I shall end this introductory text with a grand eclectic finale and bring together the Parsonian, Marxist and Mannheimian thoughts which we have collected in the various preceding chapters of this book.

I suggest that we combine Mannheim's methodology of the *principia media* with Parsons's 'theory of general social evolution' as well as with the Marxist analysis of the concrete historical process up to the present. Parsons's evolutionary paradigm, I propose, may be viewed as the 'coherent symptomatic' of Mannheim's dialectically unfolding 'history'. Thus the unifying theme underlying and connecting the discrete manifestations of human society through history is the process of functional differentiation, inclusion, value generalisation and adaptive upgrading. In this way, Parsons's evolutionary paradigm presents the *general* laws of the evolution of human societies, but we shall need the Marxist analysis of concrete historical epochs, more particularly of the present one, to identify the historically limited *principia media* in which the general evolutionary process has become cemented, and which consequently need to be removed in order to allow the *principia media* of the next historical period – which are already with us in our own time – to gain momentum and to carry the evolutionary process on to its next stage.

The identification of these historically limited *principia media* at one and the same time leads us to pinpoint the 'ideological' character of current development models, for as we shall see these models still accept these same historically limited *principia* as 'taken-for-granted' structures of their future societies.

Which, then, are these historically limited *principia media*? Without pretending this to be an exhaustive list, let me briefly explore what I see as the most important ones.

THE IDENTIFICATION OF SOCIETY WITH THE NATION STATE

In Parsons's theoretical paradigm, as we have seen before, society is that social system (that is that system of human interaction) which has relatively (that is compared to all other systems of human interaction) the highest degree of self-sufficiency. In the modern stage of social evolution, as we have also observed, society becomes coextensive with the nation state, as the nation state is a necessary companion of the *formative* period of economic development when the differentiation of the economy requires socio-political unification on a larger scale than ever before so as to permit continuity of economic relationships in the absence of primordial social ties, to permit markets large enough to accommodate specialisation and division of labour, and to provide a new commitment and sense of belonging to the uprooted peasants who are turned into an industrial proletariat. The nation state was indeed a perfect fit for the period of competitive liberalism and limited international trade between modernising societies who were all in the same formative period and roughly of equal adaptive strength. But, today, technological advance coupled with the *historical* fact of European domination and its imposition of an international division of labour *ipso facto* presents us with a degree of economic differentiation which is no longer usefully served or indeed contained by national boundaries. Economic activity, that is society's adaptive function, has definitely and absolutely outgrown the nation state, and, therefore, to cling to the myth of national sovereignty – as all development models do, socialist or capitalist – is to cling to an anachronism, a historically limited *principia media.* Capitalist and socialist development models simply do not take into account the undeniable and irreversible facts of history, and therefore they are 'ideological'.

These facts of history which we have already discussed in the second part of the book have resulted in a global division of economic activity no longer between nations but between giant corporations. Several multinationals now have a larger annual turnover than the national product of all but the very largest of nation states. General Motors' 35 billion dollars' worth of annual sales tops the gross national product of 130 countries. Some four multinationals between them (General Motors, Standard Oil, Ford, and Royal Dutch) have a larger turnover than the gross output of *the whole continent* of Africa.[7] Effectively, these multinationals have become more self-sufficient than all but the largest of nations. They allocate resources, co-ordinate production, distribute income, protect their 'political' interests (that is *vis-à-vis* other enterprises and national governments), maintain their own patterns of cultural values (often by actively supporting and promoting educational and ideological campaigns), integrate their members (that is their own staff) and even 'socialise' their members into the company's cultural ethos and rules – all these they do more effectively and more vigorously than most nation states. Has anyone ever heard of an 'area' manager of a multinational 'break-away' group fall in line with national policies concerning, for example, employment or taxation? Yet secessions from nation states by disgruntled tribes or regions are the order of the day in developing countries.

In terms of the main currents of the next historical period, the nation state is simply 'not on', and frantic attempts at 'regionalisation' of which the European Economic Community is but an example, are basically no more than a natural historical progression to evolve socio-political units which once more may become adequate integrative structures for the economic realities of the late 1970s.

INDUSTRIALISATION AND ADAPTIVE UPGRADING

All developing countries, whether capitalist or socialist, strive to obtain material standards of living comparable to those prevalent in the West. And all but the very tiniest amongst them see industrialisation as a means towards that end. Quite apart from the question of whether or not *independent* industrialisation, as the socialist model particularly advocates, is a realistic possibility in view of existing conditions of global political economy, one should assess the realisability of industrialisation for all the developing world *per se* in terms of *ecological* and *geological* conditions. Ever since the Club of Rome's report[8] sounded the first loudly heard alarm, there has been a constant flow of scientific evidence which shows that neither the world's non-renewable resources nor the world's biosphere can stand another thrust of industrialisation of the size and scale as, say, we now have in the developed world.[9] As Raymond Aron bluntly put it. 'three and a half thousand million people *cannot* consume raw materials with the same voracious appetite as the Americans.'[10]

In view of the limitations of the world's resources, industrialisation of the scope envisaged in development models is simply 'not on'. Increased global industrial activity will simply lead to an even madder scramble for the earth's diminishing resources whereby the already rich nations, including the powerful socialist states, will scoop up the dwindling minerals because they will be able to pay the increasingly higher prices for them. The case of oil is already an illustrative example. Whilst the quadrupled oil prices so far have only meant a cost of 2 per cent of the total G.D.P. of the O.E.C.D. countries, they have put many developing countries practically out of business. On the other hand, even if there were sufficient raw materials available for further global industrialisation, such industrialisation would no longer lead to further 'adaptive upgrading', but rather to increasing and cumulative damage to the world's ecosystem, damages which would require forever more technological and forever greater industrial efforts to repair and to overcome, if indeed they are repairable.[11]

At the same time, and this is the 'Catch 22' of our earth's plight, the first of the world's most pressing problems, namely galloping population growth, can only be effectively reduced by – not so much the process of industrialisation itself as the dogmatic Marxists would have us believe – but by the material benefits and security which *result* from industrialisation. For the undeniable fact of demographic trends is that a sharp fall in birth-rates only comes with dramatic improvements in people's material standards of living.

The logic of the contemporary situation therefore points in the direction not of increased global industrialisation, and certainly not of independent industrialisation in ever so many new nation states, for this would involve an inefficient and wasteful duplication of industrial efforts, but to a 'utopian' redistribution of the already produced wealth on a world-wide scale. Such redistribution of wealth would not just be a one-time occurrence but would have cumulative effects in that, by drastically reducing the standards of living of the people in the rich world, it would reduce *their* requirements for industrial production, and hence for raw materials, leaving more to be allocated to the poor world.

Needless to say, such a utopian model of development entails a most dramatic reorganisation of the world economy. In this model neither the allocation of the world's resources of energy and minerals, nor the co-ordination of the production of goods and services, nor indeed the distribution of income, would be left to the operation of a free-market system. For the market which is an effective allocator of resources when both demand and supply are unlimited ceases to 'work' in a situation where demand is unlimited and supply restricted, and where, moreover, buyers and sellers are of unequal economic strength. Rather, the organisation of economic processes will have to move into the hands of some *world 'planning' authority*. It is at this point that I am sure the reader will shrug his shoulders and say 'utopia', but in full agreement with the title and much of the content of René Dumont's latest work I would reply to that remark with 'utopia or else'.[12] René Dumont has utopian dreams about a world authority which would allocate the U.S. 6 per cent of the world population only 6 per cent of total world energy instead of the 40 per cent or more which they are using (or wasting) now. Dumont's utopia would not permit a British cat or dog to be better fed than an Indian human, nor permit motor cars to take up twenty-five times as much urban land *per capita* as public transport and cause as much of 25 per cent of our air pollution into the bargain. Dumont's utopia calls for an austerity programme that puts an end to the rich man's wastes at the poor man's expense, with the added bonus of less work and less soul-destroying types of work for all concerned. Dumont's utopia would certainly not permit a farmer in the United States to reduce his wheat production because the price is right and a bumper harvest might depress it when 150 million inhabitants of this earth are threatened by immediate starvation. Nor would Dumont's utopia allow 50 per cent of the world's grain to feed cattle which reduces the nutrition value to be extracted by a factor 16, nor would it permit 36 per cent of the world's fish to be turned into meal, almost all of which is used for animal fodder, thus reducing the amount of protein to a small proportion.

The 'world authorities' in Dumont's utopia replace the price mechanism with a whole battery of levies, taxes and duties, but Dumont remains vague and uncommitted regarding the source of legitimation upon which the executive powers of the world authorities would be based. In fact, paradoxically, he shrinks away from the idea of world government and prefers a world economy mysteriously suspended over 'national groupings'. But here is where I believe our sociological 'utopian' analysis may lead us beyond the limitations of Dumont's economic fantasies.

PLANNING FOR UTOPIA

Referring back to Parsons's 'evolutionary' *paradigm*, which we have accepted as presenting the *general* laws of social evolution, we should now simply ask ourselves which is the next stage of differentiation and which is the next form of integration? Unfortunately Parsons's own *theory* does not go beyond the modern stage. As was seen in Chapter 2, his scheme of functional requisites, namely four only, and his conception of evolution as a progressive differentiation of each of the four functions of society, cannot by definition go beyond four stages after the first and undifferentiated stage. And this, incidentally, proves his to be yet another 'ideological' rather than utopian theory.

Which, then, should we invent as the fifth functional category? Which functional requisite of society has Parsons omitted? This fifth functional category which Parsons has forgotten, but which, I hasten to add, all historical societies failed to carry out (possibly one reason why they did not survive) is the function of *planning*. Thus far, all historical societies have failed to plan for a future under altered historical conditions. Having no conception of the historical process and therefore unaware of the inevitable changes, environmental, demographic and social, which are brought about by man's and society's very own interaction with the environment, none developed the wisdom or the techniques to plan the course of society in a future changed historical period.

What, then, is 'planning'? Following Mannheim I refer to planning where man and society 'advance from the deliberate invention of single objects or institutions to the deliberate regulation and intelligent mastery of the relationship between these objects and institutions'.[18] Although ecological crises had not penetrated the human awareness in Mannheim's time, his definition none the less covers this aspect of planning too. For planning involves both the invention of social institutions and techniques to adapt to the environment, and the understanding of the subsequent alterations in the environment wrought by these very same institutions and techniques. The conception of planning as a functional requisite for society's survival is totally new, and constitutes the main *principium medium* of the next historical period.

However, in line with Parsons's paradigm of social evolution, we now have to address ourselves to the question regarding the process of structural differentiation of the planning function, and the new integrative solution consequent upon this process of structural differentiation. As was the case with the other functions of society, I propose that the structural differentiation of the function of planning also involves a relative degree of autonomy *vis-à-vis* other spheres of society, the existence of an autonomous principle and the specialisation of the planning activity into separate subcollectivities and roles. Since we are talking here about planning of the world's social system, the task of planning should be carried out by international functional groups of technocrats and social engineers. The autonomous principle, replacing the autonomous principle of the price mechanism of the previous epoch, should be the *instrument of planning*, which is the computerised information-processing of all available data on the world's demographic, ecological and geological systems and the effect of man's technological impact upon them. The Club of Rome's M.I.T. computer model of the various exponential curves of human activity and the effects of this activity upon the environment (that is exponential curves of population growth, industrialisation, urbanisation, mineral extraction and pollution) and the interrelationships amongst these various curves is an excellent example of the kind of information-processing involved in planning. Taxes, levies and duties, as in Dumont's model, could indeed be used as executive means to carry out planning.

The *integrative* solution, attendant upon this new and higher stage of social differentiation is that of a societal community which corresponds to the scope of world planning, namely a *world society*, in which the criterion of membership is no longer that of voluntary association, since the function of planning on a global scale cannot be effectively carried out if individuals and groups are permitted to 'join or not to join'. Hence the criterion of membership, in keeping with the planning function, must simply be that of 'world inhabitant'. Literally everyone who inhabits this earth automatically belongs to world society, and becomes an equal 'unit' in the mechanised and computerised planning process. Whilst this may sound like an abhorrent limitation on the individual's freedom, it automatically gives each human being the same rights and duties, and an equal participation in the adaptive wealth of world society. That, to me, seems a more humane position than that of 'citizen of a nation state' in a world that has become so divided between privileged and disadvantaged nations that the very principle of voluntary association no longer obtains *anyway* as a criterion of membership. For, unlike the formative period of modernisation when – as in the United States – immigrants were welcome to join the society on the basis of their free will, and to obtain citizenship on the strength of their voluntary submission to the Constitution, today the rich nations are firmly sealing off their borders against unwanted immigrants. Indeed, membership of the 'relevant societal community' has once more regressed to parochial, ascriptive criteria, namely of birth, descent and, for the lucky ones, marriage with nationals of the country of orientation. In this way one is persuaded to suggest that voluntary association too has become yet one more *principium medium*, reminiscent of an earlier phase in the modern stage of social evolution.

In the presentation of Parsons's evolutionary theory – in Chapter 2 – we observed how each progressive stage of social differentiation calls for *supportive* changes in the other, already differentiated, structures of society. In this way each historical period (or stage of evolution) becomes characterised by the dominance of *one*, namely the last, differentiated function, which sets the terms of reference for the other substructures. Thus, in the modern stage, as we observed, the differentiation of the adaptive function (that is the economy) logically called for a value system which stressed rationalisation of action and voluntarism, an integrative structure which provided a formal legal system, and finally a political structure characterised by democratic association. In the same manner, the planning stage of social evolution – our 'utopia' – will require supportive changes in the cultural, integrative and political spheres. To begin with we need a new value system to legitimise the new all-inclusive world order, and the definition of membership on the basis of 'world inhabitant', as well as to provide guidelines for the priorities of world planning. The realities of the late 1970s can no longer accommodate the values of voluntarism, meritocracy and, indeed, of 'individual freedom' appropriate to the 'modern' stage of evolution. The notion that each individual is his own destiny, that each man should get what he deserves as a consequence of his own hard labour, and that each man should be free to make his own choices is yet another historically limited *principium medium*. For the freedom of individuals in privileged nations of the 1970s directly threatens the livelihood of millions of people in more unfortunate lands. Our desire and free decision to eat meat, to run motor-cars – or indeed write books – adversely affects their opportunity to eat at all, to use agricultural machinery or to protect their soil from erosion. Under the structural conditions of world capitalism, freedom negates social justice and begets inhumanity. In order to protect the standards of living and the freedom of consumer choice of the peoples of the rich world, the people in the poorer nations have to become 'disposable'. And, already, intellectuals of our time are advancing rationalisations and justifications for the 'disposable' principle. For example, it is argued that the provision of massive humanitarian relief and aid to the starving millions in India and Bangladesh only leads to more people being saved who, in turn, will raise the population still higher. As a result, it is said, relief leads to more people in crisis, still greater need for relief, and eventually to a situation that relief cannot handle. By way of dramatisation of this argument, the life-boat analogy is used: if people in the boat helped too many others to get aboard, they would all sink.[14] The fundamental premise underlying this argument is once again a voluntaristic, meritocratic conception of man: those already on board derive their right to remain there on the basis of their competitive strength because they climbed aboard first. It is the 'survival of the fittest' theorem all over again. But the life-boat analogy is false, for nobody fights his way aboard the life-boat called earth, we are all *born* aboard. Rather than dispense with its surplus passengers we should dispense with wasteful class divisions and luxuries and recognise the right of *all* to remain aboard. This, indeed, requires a massive overhaul of our value system, of our voluntaristic competitive conception of man, and its replacement by a social concept of man.

Completing our sketch of 'utopia' in concrete detail goes beyond the task of this book, and beyond the imagination of the present author. Utopias are often only premature truths. As such, they are – in any one historical period – only as visible as the tip of an iceberg. Nevertheless we should study these utopias, and watch the tip of the iceberg with great care, for, as Mannheim has said, every age allows, in differently located social groups, the rise of those ideas and values in which are contained, in condensed form, the unrealised and unfulfilled tendencies which represent the needs of each age. By collating these often discrete and seemingly unconnected ideas we may put together a picture of the only kind of society that is viable in our time.

Notes and References

INTRODUCTION

1. N. A. Simms, *Opting for Development, Guide to Development Studies in British Higher Education* (London: Overseas Development Institute, 1968), quotation from back cover.
2. L. B. Pearson, *Partners in Development* (London: Pall Mall, 1970). The committee reckons, for instance, that in the fiscal year of 1969 Brazil's share of total technical assistance under the U.S. aid programmes was 137 per cent of the United States' share (p. 182).
3. André Gunder Frank, *The Sociology of Development and the Underdevelopment of Sociology* (London: Pluto Press, 1970).
4. For a list of these fifty largest corporations, see *Fortune* (Aug 1974) pp. 184–5.
5. For the distinction between development as process, and development as action, I am indebted to the instructive lectures of Professor C. A. O. van Nieuwenhuyze at the Institute of Social Studies, The Hague.

CHAPTER ONE

1. Good introductory texts on the literature of social change are, F. R. Allen, *Socio-Cultural Dynamics* (New York: Macmillan, 1971); J. A. Ponsioen, *The Analysis of Social Change Reconsidered* (The Hague: Mouton, 1969); P. S. Cohen, *Modern Social Theory* (London: Heinemann, 1968) chs 7 and 8. An excellent collection of readings, both classic and modern, on social change is A. Etzioni and E. Etzioni (eds), *Social Change* (New York: Basic Books, 1964).
2. Cf. Ponsioen, *The Analysis of Social Change Reconsidered*, 1, p. 13.
3. A. Kingsley Davis, *Human Society*, as mentioned in Allen, *Socio-Cultural Dynamics*, p. 37.
4. R. MacIver, and C. Page, *Society: an Introductory Analysis* (New York: Rinehart, 1949) as mentioned in Allen, *Socio-Cultural Dynamics*, p. 37.
5. M. Ginsberg, 'Social Change', *British Journal of Sociology*, vol. 9, no. 3 (Sep 1958) as mentioned in Allen, *Socio-Cultural Dynamics*, p. 37.
6. A classic text here is R. S. Lynd and H. M. Lynd, *Middletown in Transition* (New York: Harcourt, Brace & Co., 1937).
7. See, for example, O. Spengler, *The Decline of the West* (London: Allen & Unwin, 1926); A. Toynbee, *Study of History* (Oxford University Press, 1954); and P. Sorokin, *Social and Cultural Dynamics*, vols I–IV (London: Allen & Unwin, 1937–41).
8. A good discussion on this point is to be found in W. E. Moore, *Social Change* (Englewood Cliffs, N.J.: Prentice-Hall, 1963) especially ch. 3.

9. For a systematic discussion of this substantive theorem, see the introduction by Robert Nisbett in *Social Change*, ed. Nisbett (Oxford: Blackwell, 1972).

10. Both the substantive issues under (*c*) and (*d*) have been most adequately dealt with by those social philosophers and historians who concerned themselves with the formal subject of social change as change of entire civilisations, namely Spengler, Toynbee, Sorokin, Kroeber, Weber and Marx. For a systematic discussion of their respective positions see Allen, *Socio-Cultural Dynamics*, especially ch. 8.

11. H. Spencer, *First Principles* (London: William & Norgate, 1911) p. 358.

12. E. Durkheim, *The Division of Labor in Society* (New York: The Free Press, 1964).

13. F. Tonnies, *Gemeinschaft und Gesellschaft*; an English translation of this book was published in 1940 under the title *Fundamental Concepts of Sociology*.

14. H. Morgan, *Ancient Society*, quoted by E. Service in his contribution on 'Evolution' in *International Encyclopaedia of the Social Sciences* (New York: Macmillan, 1968) p. 223.

15. M. Ginsberg, *Essays in Sociology and Social Philosophy*, vol. III (London: Heinemann, 1961) p. 3.

16. T. Parsons, *Societies* (Englewood Cliffs, N.J.: Prentice-Hall, 1966) and *The System of Modern Societies* (Englewood Cliffs, N.J.: Prentice-Hall, 1971).

17. M. Sahlins and E. Service (eds), *Evolution and Culture* (University of Michigan Press, 1960).

18. Parsons, *The System of Modern Societies*, p. 3.

19. Sahlins and Service, *Evolution and Culture*, p. 30.

20. Parsons, *Societies*, p. 10.

21. R. N. Bellah, 'Religious Evolution', *American Sociological Review* (June 1964) p. 358.

22. Sahlins and Service, *Evolution and Culture*, p. 37.

23. Ibid. p. 74.

24. Ibid. p. 36.

25. Ibid. p. 22.

26. S. N. Eisenstadt, 'Social Differentiation, Integration and Evolution', *American Sociological Review* (June 1964) p. 376.

27. Parsons, *Societies*, p. 22.

28. Sahlins and Service, *Evolution and Culture*. These authors suggest the following distinction between specific and general evolution:

> Specific evolution is the phylogenetic, adaptive, diversifying, specialising, ramifying aspect of total evolution. It is in this respect that evolution is often equated with movement from homogeneity to heterogeneity. But general evolution is another aspect. It is the emergence of higher forms of life, regardless of particular lines of descent or historical sequences of adaptive modification. In the broader perspective of general evolution organisms are taken out of their respective lineages and grouped into types which represent the successive levels of all-round progress that evolution has brought forth (p. 16).

Talcott Parsons is less clear cut and precise in his definition of general versus specific evolution. That he is nevertheless committed to the same distinction, the following quote testifies:

> Socio-cultural evolution, like organic evolution has proceeded by variation and differentiation from simple to progressively more complex forms. Contrary to some early conceptions in the field, however, it has not proceeded in a single neatly definable line, but at every level has included a rather wide

variety of different forms and types. Nevertheless, longer perspectives make it evident that forms apparently equally viable in given stages and circumstances have not been equal in terms of their potentialities for contributing to further evolutionary developments. Still, the immense variability of human patterns of action is one of the most important facts about the human condition (*Societies*, p. 2).

29. T. Parsons, 'Evolutionary Universals', *American Sociological Review* (June 1964) especially p. 341.
30. Ibid. p. 339.
31. Sahlins and Service, *Evolution and Culture*, pp. 54–5. For Sahlins and Service, this tendency toward the stability of culture presents at the same time a *law* of the non-linear nature of progress, namely in that an advanced form does not normally beget the next stage of evolutionary advance; the next stage begins rather in a different line (pp. 98–9).
32. Ibid. p. 73.
33. Ibid. p. 90.

CHAPTER TWO

1. Parsons, *Societies* and *The System of Modern Societies* (1966 and 1971 resp.).
2. The three contributions which form a unity together were all published in the *American Sociological Review* (June 1964). Therein the articles were: T. Parsons, 'Evolutionary Universals'; R. N. Bellah, 'Religious Evolution' and S. N. Eisenstadt, 'Social Differentiation, Integration and Evolution'.
3. Cf. excerpts of their work in T. Parsons, E. Shils, K. D. Naegle and J. R. Pitts (eds), *Theories of Society*, vols I and II (Glencoe, Ill.: The Free Press, 1961).
4. Of Max Weber's work six volumes have been of special relevance to neo-evolutionary theory: first his work on comparative religions as in: *Ancient Judaism* (Glencoe, Ill.: The Free Press, 1952) and *The Religion of China* (Glencoe, Ill.: The Free Press, 1951); secondly, his work on capitalism, rationalism, bureaucracy and the modern state, as in *The Theory of Social and Economic Organisation* (Glencoe, Ill.: The Free Press, 1947) and in *General Economic History* (New York: Collier-Macmillan, 1966); thirdly his comparative work on the city and the rise of the bourgeoisie as in *The City* (New York: Collier, 1962); and finally his work on the Reformation as in *The Protestant Ethic and the Spirit of Capitalism* (New York: Scribner, 1958).
5. T. Parsons, *The Social System* (Glencoe, Ill.: The Free Press, 1951). See also his 'General Introduction' to *Theories of Society* as well as his contribution in the *International Encyclopaedia of the Social Sciences*, under the concept of 'Sytems Analysis'. The latter is perhaps the briefest and most concise statement of his position.
6. Cf. P. Sorokin, *Sociological Theories of Today* (New York: Harper International, 1966) p. 407 where he says that 'Parsons' predilection for expressing platitudinous ideas in a ponderously complicated form – and his bent for unintelligible "analytical theorizing" vitiate greatly the clarity of his ideas, the exact meanings of his terms, and the adequacy of his definitions and theories'. For still more damaging criticisms see C. Wright Mills, *The Sociological Imagination* (New York: Oxford University Press, 1959). Mills claims that Parsons's voluminous work *The Social System* can be translated into a few pages of intelligible English and benefit from it.
7. A. R. Radcliffe-Brown, 'The Study of Kinship Systems', in *Theories of Society*, ed. Parsons *et al.*, p. 279.

8. W. E. H. Stanner, 'The Dreaming', in *Reader in Comparative Religion*, ed. W. Lessa and E. Z. Vogt (Evanston, Ill.: Row & Peterson, 1958).

9. Bellah, 'Religious Evolution', pp. 362–3.

10. Ibid. pp. 366 ff.

11. Cf. M. Eliade, *The Myth of the Eternal Return* (London: Pantheon Books, 1964) especially pp. 34 ff. There is a useful categorisation of the ideal typical characteristics of primitive worldviews in ch. I, 'Archetypes and Repetition'. See also H. Frankfort and H. A. Frankfort (eds), *Before Philosophy* (Harmondsworth: Penguin, 1971); they categorise the differences between primitive and scientific world views in ch. 1, 'Myth and Reality'.

12. This argument follows R. Barrington Moore, *The Social Origins of Democracy and Dictatorship* (Harmondsworth: Penguin, 1969) especially pp. 172–3.

13. Hornell Hart, 'Social Theory and Social Change', in *Symposium in Sociological Theory*, ed. Llewellyn Gross (Evanston, Ill.: Row & Peterson, 1959) ch. 7. Hart advances the conception of the logistic curve of cultural growth, and presents graphs depicting man's increased ability to dominate nature. He shows that since 1600 logistic 'surges' have been characteristic of social change.

14. Early sociologists like Ogburn, Chapin and Morgan had already observed the essentially geometrical growth of human culture; see ibid. p. 205.

15. It is, of course, well known that in the Middle Ages most labour-saving devices were in fact made in the monasteries. See, for example, Lynn White, *The Expansion of Technology 500–1400*, Fontana Economic History of Europe, vol. I, section 4 (London: Fontana, 1971).

16. J. Needham, 'Chinese Science and Technology', in *Clerks and Craftsmen in China and the West*, ed. Needham (Cambridge University Press, 1970). For further documentation on this point see also his *Science and Civilisation in China*, vols 1–5 (Cambridge University Press, 1954–74).

17. C. Singer, *A Short History of Science to the Nineteenth Century* (Oxford: Clarendon Press, 1949) especially ch. VII.

18. This view of the role of technology as a tool of science is, naturally, not a unanimous view. For a critical view see Rupert Hall, 'The Scholar and the Craftsman' in *Critical Problems in the History of Science*, ed. M. Clagett (Wisconsin University Press, 1962).

19. R. Firth, *Elements of Social Organisation* (London: Watts, 1951) p. 142.

20. Cf. B. Malinowski, *Argonauts of the Western Pacific* (London: Routledge & Kegan Paul, 1932).

21. Cf. T. S. Epstein, *Economic Development and Social Change in South India* (Manchester University Press, 1962) especially ch. IV.

22. Quoted in N. J. Smelser, *The Sociology of Economic Life* (Englewood Cliffs, N.J.: Prentice-Hall, 1963) p. 19.

23. Quoted in D. Beetham, *Max Weber and the Theory of Modern Politics* (London: Allen & Unwin, 1974) p. 86.

24. Parsons, 'Evolutionary Universals' p. 350.

25. M. Weber, *General Economic History* (New York: Collier-Macmillan, 1966) p. 207.

26. Ibid. 25, p. 209.

27. Beetham, *Max Weber and the Theory of Modern Politics*, p. 68.

28. Weber, *General Economic History*, 25, p. 233.

29. Bellah, 'Religious Evolution', pp. 368–9.

30. M. Weber, *The Protestant Ethic and the Spirit of Capitalism* (New York: Scribner, 1956) pp. 158 ff.

31. Ibid. p. 104.

32. Ibid. p. 141.

33. Ibid. p. 172.

34. He dates it from the seventeenth century: 'The tendency toward rationalizing technology and economic relations with a view to reducing prices in relation to costs, generated in the 17th century a feverish pursuit of invention. All the inventors of the period are dominated by the object of cheapening production. The notion of perpetual motion as a source of energy is only one of many objectives of this quite universal movement.' Ibid. p. 231.

35. For example, Marion Levy takes this as the definition of modernisation in his two volumes, *Modernisation and the Structures of Societies* (Princeton University Press, 1966).

36. T. Hughes, 'Industrialisation, Economic Aspects', in *International Encyclopaedia of the Social Sciences* p. 252.

CHAPTER THREE

1. Cf. Sorokin, *Sociological Theories of Today*, p. 605; and C. C. Zimmerman, 'Contemporary Trends in Sociology', in *Readings in Contemporary American Society*, ed. J. Roucek (New York: Paterson, 1962) pp. 3–29.

2. Parsons, *Societies*, p. 111.

3. This point is made, for example, by Anthony Smith in *The Concept of Social Change* (London: Routledge & Kegan Paul, 1973) p. 43. Parsons himself, in the last paragraph of his essay 'Evolutionary Universals', suggests that historical details should be filled in later. Smith's study, incidentally, is a very careful and balanced study of the merits and the shortcomings of neo-evolutionary theory.

4. The structural-functionalist school in anthropology is older than the one in sociology, and in many ways has been its cradle. Famous names in anthropology connected with this school are Levi-Strauss, Marcel Mauss, Malinowski and Radcliffe-Brown. Talcott Parsons's work is responsible for the firm entrenchment of the functionalist school in sociology.

5. Parsons, *The Social System*, p. 167.

6. On the methodology of the concept of functional reciprocity in anthropology, see C. G. Hempel, 'The Logic of Functional Analysis', and A. W. Gouldner, 'Reciprocity and Autonomy in Functional Theory'. Both these essays appeared in *Symposium in Sociological Theory*, ed. Gross. Hempel cites an important statement by Radcliffe-Brown on the role of functional analysis as an explanatory method, be it that Radcliffe-Brown does not regard functional analysis as the only explanatory method suited for the social sciences: 'Similarly one "explanation" of a social system will be its history where we know it – the detailed account of how it came to be what it is and where it is. Another "explanation" of the same system is obtained by showing (as the functionalists attempt to do) that it is a special exemplification of laws of social physiology or social functioning. The two kinds of explanation do not conflict, but supplement one another' (p. 282).

7. N. J. Smelser, 'Towards a Theory of Modernization', in *Social Change*, ed. A. Etzioni and E. Etzioni. This is probably the most widely quoted theoretical text on modernisation. It has been reprinted many times in various readers on social change and development, sometimes under the different title of 'Mechanisms of Change and Adjustment to Changes', as in *Industrialisation and Society*, ed. B. F. Hoselitz and W. E. Moore (The Hague: Mouton, 1963).

8. Kindleberger observed the following about these international missions of experts: 'The mission brings to the underdeveloped country a notion of what a developed country is like. They observe the underdeveloped country. They subtract the latter from the former. The difference is a program. Most of the

members of the mission come from developed countries with highly articulated institutions for achieving social, economic and political ends. Ethnocentricity leads inevitably to the conclusion that the way to achieve the comparable levels of capital formation, productivity and consumption is to duplicate these institutions.' In 'The Economy of Turkey, The Economic Development of Guatemala, and Report on Cuba', *Review of Economics and Statistics*, vol. 34, no. 4 (1952) pp. 391–2.

To attest to this, the former President of the World Bank, Eugene Staley, is on record as saying that 'Economic development will only work if countries in Asia, Africa and Latin America adopt the social institutions and values of the West.' These examples of ethnocentric policy formulation and their criticisms were collected by B. F. Hoselitz in *Sociological Aspects of Economic Growth* (Glencoe, Ill.: The Free Press, 1960) p. 55. The policy formulations which Hoselitz criticises date back to the 1950s. However, since then international organisations such as the World Bank and the United Nations have hardly matured. A very penetrating critique of the western-style policies advocated by the World Bank and often insisted upon by the World Bank staff, is Theresa Hayter, *Aid as Imperialism* (Harmondsworth: Penguin, 1971). Furthermore, as Miss Hayter shows with great force, the ethnocentric policies advocated by the World Bank's staff in actual fact facilitate the continuing stranglehold of western imperialism over the underdeveloped territories. On this see also C. Payer, *The Debt Trap; the I.M.F. and the Third World* (Harmondsworth: Penguin, 1974).

9. W. W. Rostow, *The Stages of Economic Growth* – quite unnecessarily subtitled 'A non-Communist manifesto' – (Cambridge University Press, 1960); this work has no doubt been the most influential. Other important economists who have brought in social and even psychological variables into their economic development theories are A. Lewis, *The Theory of Economic Growth* (London: Allen & Unwin, 1955) and E. E. Hagen, *On the Theory of Social Change* (Homewood, Ill.: Dorsey, 1962) and *The Economics of Development* (Homewood, Ill.: Irwin, 1968). But by far the most comprehensive of all these approaches is Gunnar Myrdal *et al.*, *Asian Drama*, vols I–III (New York: Pantheon, 1968). Myrdal unwittingly proposes nothing short of total westernisation of the underdeveloped countries when he grudgingly lists the attitudinal and social shortcomings of man and society in underdeveloped countries: low levels of work discipline, punctuality and orderliness, superstitious beliefs and irrational outlook, lack of alertness, lack of adaptability, lack of ambition and general readiness for change and experiment, and so on and so forth. See vol. III, p. 1862 for an example.

10. See, for example S. N. Eisenstadt, 'Social Change and Development', in *Readings in Social Evolution and Development*, ed. S. N. Eisenstadt (Oxford: Pergamon Press, 1970). Eisenstadt gives elites an overwhelming importance in any process of modernisation, at all times and places. Relevant is also his introductory text *Modernisation: Protest and Change* (Englewood Cliffs, N.J.: Prentice-Hall, 1966) where he sees the main problem of modernisation as hingeing upon the ability of the political centre to deal with the continuous process of structural differentiation which leads to the impingement of ever broader groups on the centre. Whereas Eisenstadt's are serious studies of the role and structure of modernising elites and the stresses under which they operate, David McClelland is an example of social engineering gone too far. His study of the aptitudes and the attitudes of modernising western elites (the protestant bourgeois entrepreneurial elites) led him to the bizarre extreme of actually attempting to force the aptitudes of elites from developing countries into the same mould. He believes that elites in developing countries lack the achievement motivation so characteristic of the competitive protestants of the

western world. For a report on his training courses see D. McClelland and D. G. Winter, 'Motivating Economic Achievement', in *The Challenge of Development*, ed. R. D. Ward (Chicago: Aldine, 1967). The theoretical basis for his social engineering is to be found in an earlier work, *The Achieving Society* (New York: van Nostrand, 1961). How McClelland is illegitimately reducing the complexity of development and evolution to one single psychological factor is clearly argued in Eisenstadt's critique: see S. N. Eisenstadt 'The Need for Achievement', *Economic Development and Cultural Change*, vol. XI, 4 (1963) pp. 420–31.

11. This is generally the line taken by the economic-development theorists mentioned in note 9. But even sociologists and anthropologists have tended to accept the merits of the technocratic order of modern societies rather uncritically. Many American social scientists have adopted a civilising missionary approach to development because of this uncritical acceptance of the benefits of their own technological society. Hence there is an almost inexhaustible supply of studies concerned with the question of how to facilitate the adoption of technology in non-western countries. Classic texts of this kind are: C. M. Ahrensberg and Arthur Niehoff, *Introducing Social Change, a manual for Americans overseas* (Chicago: Aldine, 1964); A. M. Niehoff (ed.), *A Casebook of Social Change* (Chicago: Aldine, 1966); A. M. Niehoff and J. C. Anderson, 'The Process of Cross-Cultural Innovation', *International Development Review*, 6, 2 (1964) which offers an analysis of 106 published case histories of efforts to introduce innovations into developing countries. E. H. Spicer (ed.), *Human Problems in Technological Change* (Russell Sage Foundation, 1962) presents similar case histories. M. Mead (ed.), *Cultural Patterns and Technical Change* (UNESCO, 1955; reprint New York: New American Library, 1957). Margaret Mead's book studies especially the psychological tensions and illnesses associated with technical change. By far the most naively jubilant description of that super creature called modern man can be found in Alex Inkeles, 'The Modernization of Man' in *Modernization, the Dynamics of Growth*, ed. Marion Weiner (New York: Basic Books, 1966). A critique of the approach represented by these works is A. Mazrui, 'From Social Darwinism to Current Theories of Modernization', *World Politics*, XXI (Oct 1968) pp. 69–83.

12. Examples of this position are less easily available, but see Robert Redfield, *The Primitive World and its Transformations* (Cornell University Press, 1953) as a clear demonstration of this position. Redfield regrets the loss of the 'moral order' of primitive society in the face of the 'technical order' of advanced societies. Lucy Mair, *New Nations* (London: Weidenfeld & Nicolson, 1963) is another, though less clear-cut example.

13. Parsons, *The Social System*, pp. 58–67.

14. Ibid. p. 177. Note that Parsons uses only four of the five pattern variables to characterise the modern economic system. A sharp critique of the empirical reliability of these pattern variables has been made by A. G. Frank, *The Sociology of Development and the Underdevelopment of Sociology* (London: Pluto Press, 1971).

15. W. E. Moore, 'The Social Framework of Economic Development', in *Tradition, Values and Socio-Economic Development*, ed. R. Braibanti and J. J. Spengler (Durham, N.C.: Duke University Press, 1961).

16. Cf. W. J. Goode, 'Industrialisation and Family Change', in *Industrialisation and Society*, ed. Hoselitz and Moore, pp. 237 ff. See also his introductory text *The Family* (Englewood Cliffs, N.J.: Prentice-Hall, 1964).

17. Cf. T. Parsons, 'The Kinship System of the Contemporary United States', in *Essays in Sociological Theory*, ed. Parsons: rev. edn (Glencoe, Ill.: The Free Press, 1954). Parsons presents the theoretical argument. Many students of developing countries have observed the irresistible advance of both nuclear

family patterns and the romantic-love complex. For comparative studies of changing family patterns all over the world, see W. J. Goode, *World Revolution and Family Patterns* (Glencoe, Ill.: The Free Press, 1963).

18. Cf. D. H. Smith, 'Modernization and the Emergence of Voluntary Organisations', *International Journal of Comparative Sociology*, vol. 13, no. 2 (1972) and Smelser, 'Towards a Theory of Modernization'.

19. See, for instance, C. E. Black *The Dynamics of Modernisation* (New York: Harper & Row, 1967) for whom the creation of the national state is the essential feature of modernisation.

20. D. S. Landes, 'Industrialisation and the Development of Industrial Societies', in *Industrial Man*, ed. T. Burns (Harmondsworth: Penguin, 1969) p. 78.

21. W. E. Moore, 'Industrialisation: Social Aspects', in *International Encyclopaedia of the Social Sciences*, pp. 263–70. Moore argues that nominal ownership by the State merely changes the conditions of transferability of property rights in detail, as transfers of power over, and responsibility for, the materials of production are still necessary.

22. These effects on the rural social structure have been observed on all continents. See, for example, E. J. Hobsbawm, *The Age of Revolution* (London: Weidenfeld & Nicolson, 1962) ch. 8; and E. Wolf, *Peasants* (Englewood Cliffs, N.J.: Prentice-Hall, 1966). Some detailed descriptions of this process can be found in T. Shanin (ed.), *Peasants and Peasant Societies* (Harmondsworth: Penguin, 1971). See, for example, Ernest Feder, 'Latifundia and Agricultural Labour in Latin America', pp. 83–97. See also, Eric Wolf, 'The Hacienda System and Agricultural Classes in San Jose, Puerto Rico', in *Social Inequality*, ed. André Beteille (Harmondsworth: Penguin, 1969), pp. 172–90.

23. Cf. Beetham, *Max Weber and the Theory of Modern Politics*, p. 83.

24. R. Firth, *Elements of Social Organisation*, London: Watts, 1951), p. 143.

25. For good examples of this form of argumentation consult American political scientists such as L. Pye, *Aspects of Political Development* (Boston: Little, Brown & Co., 1966) pp. 72 ff.; D. Lerner, *The Passing of Traditional Society* (New York: The Free Press, 1958). Lerner sees democracy as a last development in societal growth; a 'crowning' institution of the – what he calls – 'participant' society. See also R. Dahl, *A Preface to Democratic Theory* (Chicago University Press, 1956) who argues that cross-secting, politically relevant affiliations are important for democracy because they lead to tolerance. For a good summary and a strong support for these views, see S. M. Lipset, *Political Man* (New York: Doubleday, 1960).

26. Cf. Lerner, *The Passing of Traditional Society*, which argues that modernisation hinges upon literacy for that very reason.

27. For comparative evidence on this backward-sloping supply curve see W. E. Moore, *Industrialisation and Labor* (Cornell University Press, 1951) pp. 35–7.

28. Cf. R. H. Tawney, *Religion and the Rise of Capitalism*, 2nd edn (London: Murray, 1964) pp. 269–70.

29. See G. Balandier, 'Comparative Study of Economic Motivations and Incentives in a Traditional and in a Modern Environment', General Report on the Round Table Conference organised by the International Research Office on the Social Implications of Technological Change (Paris, Mar 1954) in J. Meynaud (ed.), *Social Change and Economic Development* (UNESCO, 1963) pp. 29 ff.

30. For internationally accepted 'standard' compilations of such indicators, see, for example, *Compilation of Indicators of Development* (Geneva: UNRISD, 1969; *Contents and Measurement of Socio-Economic Development*, a staff study of the United Nations Research Institute for Social Development

(New York: Praeger, 1972) pp. 45–6; N. Ginsberg, *Atlas of Economic Development* (Chicago University Press, 1961); A. S. Banks and R. B. Texter, *A Cross-Polity Survey* (Massachusetts Institute of Technology Press, 1963); B. M. Russett *et al., World Handbook of Political and Social Indicators* (Yale University Press, 1964). C. L. Taylor and M. Hudson, *World Handbook of Political and Social Indicators*, 2nd edn (Yale University Press, 1972).

31. We shall come back to this problem of over-urbanisation in Part Two, Chapter Five. For a bibliography on over-urbanisation see A. A. Laquian, *A Selected Bibliography on Rural–Urban Migrants' Slums and Squatters in Developing Countries* (Monticello, Ill.: G.P.L., 1971). See also *Urban Agglomerations in the States of the Third World: their Political, Social and Economic Role*, report of the 34th Incidi study session held in Aix-en-Provence, 11–16 Sep 1967 (Brussels, 1971).

32. Cf. R. Jolly, E. de Kadt, H. Singer and F. Wilson (eds), *Third World Employment* (Harmondsworth: Penguin, 1973) Part Four, 'Education and Employment', especially the article by P. H. Coombs, 'The World Educational Crisis; a systems analysis', p. 194. The misfit between western formal education on the one hand and development needs on the other has been signalled for many years; see the scornful attack by R. Dumont in *False Start in Africa* (London: Andre Deutsch, 1966) ch. 7.

33. See especially Theresa Hayter's devastating critique of the World Bank's policies in *Aid as Imperialism* (Harmondsworth: Penguin, 1971).

34. That is the Harvard Business School. The United States also trains the military of developing countries, especially of Latin American countries, in capitalist development strategies and techniques. Reference to these superior 'war colleges' is made by A. Stephen, *The Military in Politics, Changing Patterns in Brazil* (Princeton University Press, 1971) p. 217.

35. For a revealing account of these C.I.A. activities in Third World countries, consult D. Horowitz, *From Yalta to Vietnam* (Harmondsworth: Penguin, 1967). The *Guardian* (26 September 1974) reports that Dr Kissinger in defending the C.I.A. activities in Chile during the Allende regime, had said that U.S. activities were aimed at 'keeping opposition parties in existence'.

CHAPTER FOUR

1. A. G. Frank, *Sociology of Development and the Underdevelopment of Sociology* (London: Pluto Press, 1971) p. 41.

2. Ibid. pp. 42–3.

3. K. Marx, *Capital*, especially chapters from the third volume. But his leading work on the subject are his articles: 'The British Rule in India', 'The Government of India', and others which he wrote for the *New York Daily Tribune* in 1853. For a collection of all his work on colonialism, see K. Marx and F. Engels, *On Colonialism*, edited by the Foreign Languages Publishing House in Moscow (London: Lawrence & Wishart, 1960).

4. V. I. Lenin, *Imperialism, the Highest Stage of Capitalism* (London: Lawrence & Wishart, 1916); reprinted Moscow, Foreign Languages Publishing House, 1920. This work is especially relevant to an understanding of the development of underdevelopment in the period of colonialism.

5. P. A. Baran, *The Political Economy of Growth*, 4th edn (New York: Monthly Review Press, 1967); see especially ch. v: 'On the Roots of Backwardness'.

6. A. G. Frank, *Capitalism and Underdevelopment in Latin America* (New York: Monthly Review Press, 1967) reprinted by Penguin in 1971. His theory of capitalist underdevelopment is set out in the first chapter, and is later applied

to Chile and Brazil. This makes for cross-continental comparability with Paul Baran's work, because the latter's theory of the development of underdevelopment is based on a study of India. A useful Marxist study of the development of underdevelopment in Africa is W. Rodney, *How Europe Underdeveloped Africa* (Dar es Salaam: Tanzania Publishing House, and London: Bogle L'Ouverture, 1972).

7. K. Griffin, *Underdevelopment in Spanish America* (London: Allen & Unwin, 1969) p. 34.

8. Ibid. p. 35.

9. For an entertaining discussion of these essential instruments of European overseas expansion, see C. M. Cipolla, *European Culture and Overseas Expansion* (Harmondsworth: Penguin, 1970).

10. R. Mukherjee, *The Rise and Fall of the East India Company*, 2nd edn (Berlin: Verlag, 1958) p. 36.

11. K. Marx, *Capital*, vol. I, quoted in Baran, *The Political Economy of Growth*, p. 142.

12. Baran, *The Political Economy of Growth*, p. 142. Furthermore, this claim is confirmed in W. E. Minchinton (ed.), *The Growth of English Overseas Trade in the 17th and 18th Centuries* (London: Methuen, 1969). In his own introduction, Minchinton observes: 'Merchants commonly used some of their profits to extend their scale of operations, but since the benefits of size were limited, they tended also to diversify their activities. Foreign Trade accordingly provided a surplus to finance industrial expansion', pp. 46–7.

13. Cf. W. W. Rostow, *The Stages of Economic Growth* (Cambridge University Press, 1969) p. 41.

14. Cf. G. Pendle, *A History of Latin America* (Harmondsworth: Penguin, 1971) p. 66.

15. Cf. H. E. Fisher, 'Anglo–Portuguese Trade, 1700–1770' in *The Growth of English Overseas Trade in the 17th and 18th Centuries*, ed. Minchinton, pp. 144–64. Not once in British–Portuguese trade between 1698 and 1775 was there a deficit for Britain.

16. Cf. Mukherjee, *The Rise and Fall of the East India Company*, p. 398; and Baran, *The Political Economy of Growth*, p. 146.

17. Brook Adams, 'The Law of Civilisation and Decay', quoted in Baran, *The Political Economy of Growth*, p. 146. India was not the only exogenous contributor to European capital. Eric Williams, *Capitalism and Slavery* (London: Andre Deutsch, 1972) claims this doubtful honour for the African slave trade: the profits from the West African and West Indian slave trade 'provided one of the mainstreams of accumulation of capital in England which financed the Industrial Revolution', p. 52. Others have made similar claims for the South American loot that found its way into 'the merchant coffers of Protestant Europe . . . through smuggling on a grand scale combined with piracy'. In M. Barrat Brown, *After Imperialism*, 3rd edn (London: Heinemann, 1973) pp. 32—3.

18. See P. D. Curtin, *The Atlantic Slave Trade; A Census* (University of Wisconsin Press, 1969) table i, p. 5.

19. Griffin, *Underdevelopment in Spanish America*, p. 45.

20. Quoted in ibid. p. 47.

21. Cf. A. Moorehead, *The Fatal Impact* (London: Hamilton, 1966) part I on the decimation of the population of the South Pacific Islands through the introduction of alien diseases. See also Kingsley Davis, *The Population of India and Pakistan* (Princeton University Press, 1951) ch. 6.

22. Cf. E. Durkheim, *The Division of Labour in Society* (London: Macmillan, 1933).

23. Rodney, *How Europe Underdeveloped Africa*, p. 115.

24. Ibid. p. 114.

25. Apart from the already mentioned works of Kark Marx, Paul Baran, R. Mukherjee and Brook Adams, special reference can also be made to V. Anstey, *The Economic Development of India*, 4th edn (London: Longmans Green, 1952); R. Palme Dutt, *India Today* (London: Gollancz, 1940) and R. Dutt, *The Economic History of India under Early British Rule from 1757 to 1837*, 3rd edn (London: Kegan Paul, 1901).

26. Marx, *Capital*, quoted in M. Barrat Brown, 'A Critique of Marxist Theories of Imperialism' in *Studies in the Theory of Imperialism*, ed. Roger Owen and Bob Sutcliffe (London: Longman, 1972) pp. 45–6.

27. D. C. M. Platt, *Latin America and British Trade, 1806–1914* (London: Black, 1972) pp. 14–15. Furthermore, Platt observes that 'Lancashire was never able entirely to displace the hand-loom weaver; the hand-loom industry was still supplying at least 25% of the cloth consumed in India at the beginning of the twentieth century', p. 13.

28. Ibid. pp. 13–14.

29. Cf. J. Hatch, *Nigeria: A History* (London: Secker & Warburg, 1971) p. 86.

30. Popularisation of African pre-colonial history is owed to Basil Davidson. See, for example, his *Africa: History of a Continent* (London: Weidenfeld & Nicolson, 1966) or the revised edition, *Africa in History* (London: Weidenfeld & Nicolson, 1968). His *The African Past* (Harmondsworth: Penguin, 1964) gives a collection of chronicles and documents from antiquity to the contact with Europe. See also R. Oliver and J. D. Fage, *A Short History of Africa* (Harmondsworth: Penguin, 1962).

31. Lenin, *Imperialism, the Highest Stage of Capitalism*, p. 145.

32. Ibid. p. 131. The semi-colonial character of Latin American states in the nineteenth century, that is after their independence from Spain and Portugal in precisely this aspect, is clearly demonstrated in Gunder Frank's work, *Capitalism and Underdevelopment in Latin America*, pp. 92 ff., where he quotes, *in extenso*, data from the Chilean scholar Ramirez which show the extent to which Chilean productive wealth was in foreign capitalist hands: 'Ever since the monopoly of the English foundries made them into arbiters of the price of this product (i.e. copper) and ever since they limit or extend our mining by means of their capital, the real wealth of our society remains subject to the interests of foreign speculators who, looking out for themselves, place us in the sad situation in which we are', p. 95. Incidentally, the terminology used here, of 'centre' versus 'periphery', was first used by Raoul Prebisch, Latin American scholar and Secretary-General of the United Nations Conference on Trade and Development.

33. H. Magdoff, *The Age of Imperialism* (New York: Monthly Review Press, 1969).

34. J. de Castro, *The Geography of Hunger* (Boston: Little, Brown & Co., 1952).

35. L. Pearson *et al.*, *Partners in Development* (London: Pall Mall, 1970) p. 81.

36. Marx, 'On Colonialism', quoted in M. Barrat Brown, 'A Critique of Marxist Theories of Imperialism', in *Studies in the Theory of Imperialism*, ed. Owen and Sutcliffe, p. 46.

37. Cf. I. Illich, 'Outwitting the Developed Countries', in *New York Review of Books* (6 Nov 1969); reprinted in *Underdevelopment and Development; the Third World Today*, ed. H. Bernstein (Harmondsworth: Penguin, 1973) pp. 357 ff.

38. R. Prebisch, *Towards a New Trade Policy for Latin America*, Report by the Secretary-General of UNCTAD (United Nations, 1964).

39. In Britain, for example, the Haslemere group; the World Development Movement, Third World First and the World Poverty Action Group.

40. Such proposals were, for example, made by the U.N. Economic and Social Council, Committee for Development Planning (the so-called Tinbergen Committee) and by the equally authoritative Pearson Committee in *its* study of the state of the development situation, which had been commissioned by the United Nations. See Pearson, *Partners in Development*, pp. 81 ff.

41. Cf. W. A. Lewis, 'Aspects of Tropical Trade' and T. Morgan, 'Trends in the Terms of Trade and their Repercussions on Primary Producers'; both are quoted in Angus Hone, 'The Primary Commodities Boom', *New Left Review*, 81 (Sep–Oct 1973) p. 91.

42. Cf. P. Jalee, *The Third World in World Economy* (New York: Monthly Review Press, 1969) translated by M. Klopper, p. 73. See also P. Streeten, 'The Kind of Self-help Poor Nations Need', *New Society* (13 Apr 1972) p. 60.

43. *I.M.F. Survey* (5 Aug 1974) pp. 246–7.

44. Each of these four points were also made by H. Singer, 'The Commodity Boom and Developing Countries', *New Society* (30 Aug 1973).

45. A. Emmanuel, 'Current Myths of Development', *New Left Review*, 85 (May–June 1974) p. 65.

46. H. B. Chenery, 'Growth and Structural Change', *Finance and Development*, vol. 8, no. 3 (Sep 1971) pp. 25–6. Bill Warren (see n. 47) largely bases his argument on Chenery's statistics.

47. Four articles contributed to this polemic: Bill Warren, 'Imperialism and Capitalist Industrialization', and Angus Hone, 'The Primary Commodities Boom', *New Left Review*, 81 (Sep–Oct 1973). Their positions were countered by A. Emmanuel, 'Current Myths of Development', and P. McMichael, J. Petras and R. Rhodes, 'Industry in the Third World', *New Left Review*, 85 (May–June 1974).

48. Warren, 'Imperialism and Capitalist Industrialization', p. 10.

49. Ibid. p. 24.

50. Chenery, 'Growth and Structural Change'.

51. This point is made by both groups of critics of Warren's view: Emmanuel, 'Current Myths of Development', p. 64, and McMichael *et al.*, 'Industry in the Third World', p. 101.

52. Warren, 'Imperialism and Capitalist Industrialization', p. 7.

53. McMichael *et al.*, 'Industry in the Third World', p. 101. The authors quote a study by Sofia Medez Villareal who estimates that between 1965 and 1975 Mexico will have experienced a net *loss* of industrial employment of 1,861,000 workers.

54. In ibid. p. 103.

55. Emmanuel, 'Current Myths of Development', pp. 66–7. The other critics, McMichael *et al.*, 'Industry in the Third World', p. 97, make the same point.

56. Ibid. pp. 66–7.

57. McMichael *et al.*, 'Industry in the Third World', p. 99.

58. Emmanuel, 'Current Myths of Development', p. 77.

59. M. Barrat Brown, *The Economics of Imperialism* (Harmondsworth: Penguin, 1974) p. 207.

60. R. Vernon, *Sovereignty at Bay* (Harmondsworth: Penguin, 1973) p. 71.

61. H. Magdoff, 'Imperialism Without Colonies', in *Studies in the Theory of Imperialism*, ed. R. Owen and B. Sutcliffe (London: Longmans, 1972) pp. 153–154. The same point has been made by others, for example M. Barrat Brown's chapter on neo-colonialism in *The Economics of Imperialism*, especially p. 268. See also P. A. Baran and P. M. Sweezy, *Monopoly Capital* (New York: Monthly Review Press, 1966; and Harmondsworth: Penguin, 1970) p. 111; and P. M. Sweezy and H. Magdoff: *The Dynamics of U.S. Capitalism* (New

York: Monthly Review Press, 1972) – reprint article on 'Foreign Investment', pp. 31–42.

62. McMichael *et al.*, 'Industry in the Third World', p. 86.

63. Frank, *Capitalism and Underdevelopment in Latin America*, p. 334.

64. C. V. Vaitsos, 'Bargaining and the Distribution of Returns in the Purchase of Technology by Developing Countries', in *Underdevelopment and Development*, ed. H. Bernstein (Harmondsworth: Penguin, 1973) pp. 315–22, this quote on p. 319.

65. C. Tugendhat, *The Multinationals* (Harmondsworth: Penguin, 1973) p. 170.

66. Ibid. p. 137.

67. Magdoff, 'Imperialism Without Colonies', in *Studies in the Theory of Imperialism*, ed. Owen and Sutcliffe, p. 152.

68. For an overview of such multinational tactics and strategies, Tugendhat's *The Multinationals* makes invaluable reading, especially part 2, 'How Multinational Business Works Today'. See also, Baran and Sweezy, *Monopoly Capital*, especially ch. 2, 'The Giant Corporation'; and Sweezy and Magdoff, *The Dynamics of U.S. Capitalism*, reprint article on 'The Multinational Corporation', pp. 88–112.

69. This is what Magdoff sees as the essence of the new imperialism. See 'Imperialism Without Colonies', in *Studies in the Theory of Imperialism*, ed. Owen and Sutcliffe, p. 157.

70. C. V. Vaitsos, 'Patents Revisited, Their Function in Developing Countries', *Journal of Development Studies*, vol. 9, no. 1 (Oct 1972) pp. 72–97.

CHAPTER FIVE

1. R. Delavignette, *Freedom and Authority in French West Africa* (London: Frank Cass, 1968) p. 102.

2. J. Boeke, *Economics and Economic Policy of Dual Societies* (New York: Institute of Pacific Relations, 1953) p. 103.

3. C. Furtado, 'Elements of a Theory of Underdevelopment', in *Underdevelopment and Development*, ed. Bernstein, pp. 35 and 36.

4. Delavignette estimates that between 1920 and 1930 nearly 189,000 able-bodied men, in the prime of their life, and representing the best twentieth of the population of the territory of French West Africa, were 'torn from their families, from marriage, from the village and from the fields', in *Freedom and Authority in French West Africa*, p. 113.

5. J. Nyerere, 'The Arusha Declaration', in *Freedom and Socialism: Uhuru na Ujamaa* (Oxford University Press, 1968) pp. 242–3, reprinted under the title 'Those who pay the bill', in *Peasants and Peasant Societies*, ed. T. Shanin (Harmondsworth: Penguin, 1971) pp. 375–6.

6. Frank, *Capitalism and Underdevelopment in Latin America*, pp. 46–7.

7. Hoselitz, *Sociological Aspects of Economic Growth*, p. 192.

8. W. McCord, 'Portrait of Transitional Man', in *The New Sociology*, ed. I. L. Horowitz (New York: Oxford University Press, 1964).

9. Reported in *West Africa* (15 April 1974).

10. Cf. J. Turner, 'Barriers and Channels for Housing development in Modernising Countries', in *Peasants in Cities*, ed. W. Mangin (Boston: Houghton Mifflin, 1970).

11. A. and E. Leeds, 'Brasil and the Myth of Urban Rurality, Urban Experience, Work and Values in Squatments in Rio de Janeiro and Lima', in *City and Country in the Third World*, ed. A. J. Field (Cambridge, Mass.: Schenckman, 1970).

12. F. Fanon, *The Wretched of the Earth*, 2nd edn (Harmondsworth: Penguin, 1969) p. 103.

13. E. Wolf, *Peasants* (Englewood Cliffs, N.J.: Prentice-Hall, 1966) especially pp. 50–5.

14. Ibid. p. 52.

15. R. Palme Dutt, *India Today* (Modern India Press, 1970); cf. especially p. 230 where he notes that 'many of the old traditional Zamindar families who carried on the old method of showing some consideration and relaxation for the peasants in times of difficulty broke down under the burden and were at once ruthlessly sold out, their estates being put up for auction.... A new type of sharks and rapacious business men came forward to take over the estates, who were ready to stick at nothing to extract the last anna from the peasantry in order to pay their quota and fill their own pockets.'

16. Ibid. p. 233 (quoting Lord William Bentinck's speech on 8 November 1929).

17. P. Baran, 'On the Political Economy of Backwardness', in *Imperialism and Underdevelopment*, ed. R. I. Rhodes (New York: Monthly Review Press, 1970) pp. 285–301, this quote p. 286.

18. Rodney, *How Europe Underdeveloped Africa*, p. 249.

19. Cf. M. Crowder and Obare Ikime (eds), *West African Chiefs, their Changing Status under Colonial Rule and Independence* (New York: Africa Publishing Corporation, and Ife University Press, 1970). In their introduction the editors note that 'Where chiefs were those who would have claims to rule in pre-colonial times, these same rulers had removed for them by the colonial regimes many of the limitations to their authority from below', p. xiv. See also, for the practices and policies in East Africa, D. Anthony Low and R. Cranford Pratt, *Buganda and British Overrule – 1900–1955* (Oxford University Press, 1960) p. 175.

20. Cf. Delavignette, *Freedom and Authority*, p. 85.

21. Cf. R. Anstey, *King Leopold's Legacy* (Oxford University Press, 1966) p. 208.

22. Rodney, *How Europe Underdeveloped Africa*, p. 247.

23. Sir Andrew Cohen, *British Rule in Changing Africa* (London: Routledge & Kegan Paul, 1959) p. 23, where he observes that for this purpose colonial governments often employed anthropologists.

24. Cf. K. Turner, 'The Overseas Chinese of South East Asia, National Integration and Alien Minorities', in *The Politics of New States*, ed. R. Scott, Ian Grosart *et al.* (London: Allen & Unwin, 1970) pp. 84–111.

CHAPTER SIX

1. Sir Andrew Cohen, *British Policy in Changing Africa* (London: Routledge & Kegan Paul, 1959) pp. 7, 8 and 10.

2. Delavignette, *Freedom and Authority in French West Africa*, p. 22.

3. F. W. Riggs, *Administration in Developing Countries, the Theory of the Prismatic Society* (Boston: Houghton Mifflin, 1964) especially ch. 1.

4. Ibid. ch. 3.

5. H. Frankfort and H. A. Frankfort, 'Myth and Reality', in *Before Philosophy*, ed. Frankfort and Frankfort, pp. 11–36, especially p. 13.

6. For a useful and systematic survey of magic and all its uses, see R. Firth, *Human Types, an Introduction to Social Anthropology* (London: Sphere Books, 1970) ch. VI.

7. Parsons, *Societies*, p. 93.

8. Quoted in G. Jahoda, 'Social Aspirations, Magic and Witchcraft in Ghana',

in *The New Elites of Tropical Africa*, ed. P. C. Lloyd (Oxford University Press, 1966) p. 205.

9. Ibid. p. 205.

10. Ibid. p. 206 and P. C. Lloyd, *Africa in Social Change*, rev. edn (Harmondsworth: Penguin, 1969) p. 249.

11. Riggs, *Administration in Developing Countries*, p. 176.

12. Instructive examples of these incompatibilities are to be found in W. R. Bascom and M. J. Herskovitz (eds), *Continuity and Change in African Cultures* (Chicago University Press, 1962).

13. Based upon the writer's own observations in West Africa. For a description of the secret society of the Poro amongst the Mende in Sierra Leone, see K. Little, *The Mende of Sierra Leone*, rev. edn (London: Routledge & Kegan Paul, 1967) especially ch. XII, p. 242. Little gives several traditional explanations of the Poro, none of which seem to support my thesis that the societies became secret as a result of contact with the West. The explanation closest to my interpretation is that the society originated out of men banding themselves together in secret to obviate slave-raiding parties from which they hid in the bush.

14. P. Mayer, 'Witches', in *Witchcraft and Sorcery*, ed. M. Marwick (Harmondsworth: Penguin, 1970) p. 46.

15. Most of this anthropological evidence has, however, been collected in Africa. See for instance: B. E. Ward, 'Some Observations on Religious Cults in Ashanti', in *Africa*, vol. 26 (1956) pp. 47–60; M. Marwick, 'The Continuance of Witchcraft Beliefs', in *Africa in Transition*, ed. P. Smith (London: Reinhardt, 1958); M. Marwick, 'The Sociology of Sorcery in a Central African Tribe', in *Magic, Witchcraft and Curing*, ed. J. Middleton (New York: Natural History Press, 1967); G. Jahoda, 'Social Aspirations, Magic and Witchcraft in Ghana', in *The New Elites of Tropical Africa*, ed. Lloyd; J. Beattie, 'Ritual and Social Change', *Man*, 1 (1966) pp. 60–74; V. R. Dorjahn, 'Some Aspects of Temne Divination', in *Sierra Leone Bulletin of Religion*, 4 (1962) pp. 1–9; M. Ruel, 'Witchcraft, Morality and Doubt', *ODU*, 2, 1 (1965) pp. 3–27; G. E. Simpson, 'Selected Yoruba Ritual', *Nigerian Journal of Economic and Social Studies*, 7, 3 (1965) pp. 311–24; D. Tait, 'A Sorcery Hunt in Dagomba', *Africa*, 33 (1963) pp. 136–47. For comparative evidence from North America, see C. Kluckhohn, *Navaho Witchcraft* (Boston: Beacon Press, 1967); from South America, see R. Redfield, *The Folk Culture of Yucatan* (Chicago University Press, 1941). T. Scarlett Epstein provides some evidence of the same hypothesis from India in 'A Sociological Analysis of Witch Beliefs in a Mysore Village', in *Magic, Witchcraft and Curing*, ed. Middleton.

16. Redfield, *The Folk Culture of Yucatan*.

17. Kluckhohn, *Navaho Witchcraft*.

18. Mayer, 'Witches', in *Witchcraft and Sorcery*, ed. Marwick, p. 53.

19. Marwick, 'Witchcraft as Social Strain Gauge', in ibid. pp. 280–95.

20. S. F. Nadel, 'Witchcraft in Four African Societies', *American Anthropologist*, vol. 54 (1952).

21. Marwick, 'The Sociology of Sorcery in a Central African Tribe' in *Magic, Witchcraft and Curing*, ed. Middleton.

22. T. Scarlett Epstein, 'A Sociological Analysis of Witchcraft Beliefs in a Mysore Village', in *Magic, Witchcraft and Curing*, ed. Middleton, pp. 134–54.

23. Mayer, 'Witches', in *Witchcraft and Sorcery*, ed. Marwick, p. 59.

24. H. R. Trevor Roper, 'The European Witchcraze', in ibid. pp. 121–50.

25. Mayer, 'Witches', in ibid. pp. 60–1.

26. Ibid. p. 63.

27. M. J. Field, *Search for Security* (London: Faber & Faber, 1960); quoted in Lloyd, *Africa in Social Change*, p. 255.

28. Lloyd, *Africa in Social Change*, p. 258. He also reports eight of such congregations in Ibadan in 1940, twenty-one in 1950 and eighty-three in 1962.

29. See P. Worsley, *The Trumpet Shall Sound* (London: Paladin, 1970) p. 229 where the author refers to Ronald Knox who apparently resurrected the term 'enthusiasm' in connection with cults.

30. Ibid. p. 233.

31. Ibid. p. 256.

32. Ibid. p. 239.

33. Cf. Riggs, *Administration in Developing Countries, the Theory of the Prismatic Society.*

34. Cf. J. S. Furnivall, *Colonial Policy and Practice* (Cambridge University Press, 1948).

35. Cf. K. W. Deutsch, *Nationalism and Social Communication* (New York: Wiley, 1953).

36. Riggs, *Administration in Developing Countries*, pp. 159–60.

37. For a review of the literature on the role of voluntary associations in West Africa see Ruth Simms, *Urbanisation in West Africa, a Review of Current Literature* (Evanston: North Western University Press, 1965). The writers most closely associated with the study of voluntary associations in newly emerging nations are K. Little, *West African Urbanisation, A Study of Voluntary Associations in Social Change* (Cambridge University Press, 1965) and M. Banton, *West African City, A Study of Tribal Life in Freetown* (Oxford University Press, 1957).

38. Lloyd, *Africa in Social Change*, pp. 196–202.

39. W. Mangin, *Peasants in Cities*, Readings in The Anthropology of Urbanisation (Boston: Houghton Mifflin, 1970).

40. L. Doughty, 'Behind the Back of the City: Provincial Life in Lima, Peru', in *Peasants in Cities*, ed. Mangin, p. 33.

41. For example, the continuing role of caste and kin relationships in industrialising India is stressed by R. N. Seth, *An Indian Factory, Aspects of its Social Framework* (Manchester University Press, 1968).

42. Riggs, *Administration in Developing Countries*, p. 169.

43. It is customary for hospitals in Africa to allow the families of patients to attend to and feed the patients inside the hospitals all day, and sometimes even during the night.

44. This point is particularly made by N. J. Smelser in *Essays in Sociological Explanation* (Englewood Cliffs, N.J.: Prentice-Hall, 1968). See especially the essay on 'Social Structure and Economic Development', p. 155.

45. Riggs, *Administration in Developing Countries*, p. 128.

46. Ibid. p. 171.

47. Ibid. p. 171.

48. G. Myrdal, *The Challenge of World Poverty* (Harmondsworth: Penguin, 1970) p. 239.

49. J. C. Scott, 'The Analysis of Corruption in Developing Nations', *Comparative Studies in Society and History*, vol. 11, no. 3 (1969) p. 318.

50. R. Wraith and E. Simkins, *Corruption in Developing Nations* (London: Allen & Unwin, 1963) p. 19.

51. Scott, 'The Analysis of Corruption in Developing Nations', pp. 322–3 (emphasis added).

52. J. C. Scott, *Comparative Political Corruption* (Englewood Cliffs, N.J.: Prentice-Hall, 1972) p. 35.

53. S. P. Huntington, *Political Order in Changing Societies*, 2nd edn (Yale University Press, 1969) pp. 63–4.

54. Whereas earlier there were only compilations of findings and journalistic

accounts, some major scientific work has been directed recently upon the study of the arms trade. See, for example, Stockholm International Peace Research Institute, *The Arms Trade with the Third World* (Uppsala: Almqvist & Wiksell, 1971); J. Stanley and M. Pearton, *The International Trade in Arms*, published on behalf of the Institute for strategic studies (London: Chatto & Windus, 1972); and A. C. Leiss with G. Kemp *et al.*, *Arms Transfers to Less Developed Countries*, Arms Control Project, Center for International Studies, Report C/70–1 (Cambridge, Mass.: M.I.T., Feb 1970).

55. Wraith and Simkins, *Corruption in Developing Nations*, p. 74.

56. J. C. Scott, *Comparative Political Corruption* (Englewood Cliffs, N.J.: Prentice-Hall, 1972).

57. Ibid.

58. Ibid. p. 65.

59. Ibid. p. 61, quote from Wilson.

60. Riggs, *Administration in Developing Countries*, p. 45.

61. M. McMullan, 'A Theory of Corruption', *Social Review*, 9 (1961) pp. 181–201, this quote on p. 189.

62. Wraith and Simkins, *Corruption in Developing Nations*, p. 39.

63. Ibid. p. 41.

64. Chinua Achebe, *No Longer at Ease* (London: Heinemann, 1960).

65. For a summary article on the economics of corruption, see M. J. Sharpston, 'The Economics of Corruption', *New Society* (26 Nov 1970) pp. 944–6.

66. Huntington, *Political Order in Changing Societies*, pp. 12–24.

67. Fanon, *The Wretched of the Earth*, p. 137.

68. Quoted in D. A. Rustow, *A World of Nations, Problems of Political Modernisation* (Washington, D.C.: The Brookings Institution, 1967) p. 35.

69. S. M. Lipset, *The First New Nation* (London: Heinemann, 1964) p. 16.

70. Ibid. p. 17.

71. Riggs, *Administration in Developing Countries*, p. 179.

72. Ibid. p. 180.

73. Lipset, *The First New Nation*, pp. 16–23.

74. Cf. ibid. p. 18.

75. On the role of the military in developing countries see, for example, M. Janowitz, *The Military in the Political Development of New Nations* (University of Chicago Press, 1964) and J. J. Johnson (ed.), *The Role of the Military in Underdeveloped Countries* (Princeton University Press, 1962). Both these authors stress the organisational format of the military in its capacity to intervene in domestic politics. A more recent text on the role of the military in African politics is Ruth First, *The Barrel of a Gun* (Harmondsworth: Penguin, 1970) which contains detailed case histories of several army interventions in Africa.

76. Huntington, *Political Order in Changing Societies*; Pye, *Aspects of Political Development*; D. Apter, *The Politics of Modernization* (University of Chicago Press, 1965); and Eisenstadt, *Modernization, Protest and Change*.

77. See also the report of the Committee on Comparative Politics, *Politics in Developing Areas*, ed. G. Almond and J. Coleman (Princeton University Press, 1960). The latter is probably the most frequently quoted text in this field.

78. Huntington, *Political Order in Changing Societies*, p. 3.

79. Exact source of quotation is lost, but see ibid. pp. 39–59 for the original formulation of the modernisation and violence thesis.

80. Pye, *Aspects of Political Development*, p. 8.

81. Deutsch, *Nationalism and Social Communication*.

82. Cf. Huntington, *Political Order in Changing Societies*, p. 55.

83. Arnold Toynbee, quoted in Horowitz, *From Yalta to Vietnam*, pp. 14–15.

84. Horowitz, *From Yalta to Vietnam*, p. 93.

85. C. L. Taylor and M. C. Hudson, *World Handbook of Political and Social Indicators*, 2nd edn (Yale University Press, 1972) pp. 128 and 150.

86. Ibid. calculated on the basis of data presented in tables 3.7 and 3.10.

87. *Handbook for Africa South of the Sahara* (London: Europa Publications, 1974); cf. Ruth First's introduction on 'Political and Social Problems of Development', pp. 17–25.

88. On the link between U.S. imperialist aggression on the one hand and its paranoiac philosophy of 'containing conspiracies' on the other, see Horowitz, *From Yalta to Vietnam*, especially part II. The link between U.S. aggressive foreign policy and its imperialist economic designs is clearly and persuasively argued by Magdoff in *The Age of Imperialism*, pp. 12 ff.

89. Reported in the *New Internationalist* (Sep 1974) p. 12.

CHAPTER SEVEN

1. Fidel Castro, 'History will Absolve me', see excerpt of this speech in P. E. Sigmund, *The Ideologies of the Developing Nations* (New York: Praeger, 1964) pp. 255–60, this quote p. 260. This book is a useful collection of development ideologies, as formulated by leaders of developing nations, from all three continents.

2. O. R. Lange, *Political Economy* (Oxford: Pergamon Press, 1963–70) vol. I, p. 177 (first two quotes) and p. 180 (third quote).

3. P. M. Sweezy, *Capitalist Development* (New York: Monthly Review Press, 1968) pp. 243–4.

4. Chenery, 'Growth and Structural Change', in *Finance and Development*, vol. 8, no. 3 (Sep 1971).

5. The concept of 'economic surplus', as well as the here used categorisation of capitalist versus socialist modes of utilisation of this surplus owes to Baran, *The Political Economy of Growth*, especially ch. 2.

6. For a critical discussion of the 'low-level equilibrium trap' theory, and its logical corollary the 'critical-minimum effort' theory, see H. Myint, *The Economics of the Developing Countries* (London: Hutchinson, 1969) pp. 103–9.

7. This statement is attributed to Keynes by Louis Turner, *Multinational Companies and the Third World* (London: Allen Lane, 1974) p. 3.

8. These observations were originally inspired by Baran's 'Towards a Morphology of Backwardness', in *The Political Economy of Growth*, chs I and II.

9. The failure to achieve national capitalist development via industrialisation by import substitution has been carefully researched, amongst others, by C. Furtado, *Economic Development in Latin America* (Cambridge University Press, 1970) especially chs 11 and 12. See also Santiago Macario, 'Protectionism and Industrialization in Latin America', *Economic Bulletin for Latin America*, vol. IX, no. 1 (Mar 1965); and J. H. Power, 'Import Substitution as an Industrialization Strategy', *Phillippine Economic Journal*, vol. V, no. 2 (1966). Excerpts of both these articles are reprinted in *Leading Issues in Economic Development*, ed. G. Meyer, 2nd edn (Oxford University Press, 1970) pp. 520–33.

10. The notion of stable capital–output ratios is probably the most sacred of all sacred cows in development economics. It was first developed by Harrod and Domar, and later faithfully and persuasively used by such famous (capitalist) economic growth theorists, as W. W. Rostow, *The Stages of Economic Growth* and W. A. Lewis, *Theory of Economic Growth*. For a brief summary of these growth theories see Myint, *The Economics of Developing Countries*, pp. 90–101. A critical review of the application of the concept of the capital–output ratio in contemporary developing countries comes from

P. Streeten, *The Frontiers of Development Studies* (London: Macmillan, 1972) ch. 6.

11. J. A. Schumpeter, *The Theory of Economic Development* (Cambridge, Mass.: 1934).

12. F. Stewart, 'Choice of Technique in Developing Countries', *Journal of Development Studies*, vol. 9, no. 1 (Oct 1972) pp. 99–121, especially p. 109.

13. Cf. Myint, *The Economics of Developing Countries*, especially pp. 136–142. Myint presents the run-of-the-mill argument in favour of the capital-intensive approach. For a critical appraisal of such conservative dogmas, see P. Streeten, 'Conflicts Between Output and Employment Objectives in Developing Countries', in *The Frontiers of Development Studies*, ed. Streeten. The conflict between employment and output is perhaps the most critical of currently debated economic-development issues. Recent economic theorising seems to be tilting over towards the employment argument. An authoritative International Labour Organisation report, *Employment, Income and Equality* (Geneva, 1972) has probably contributed most to this change of heart. Propagators of the employment approach are the influential scholars of the Sussex Institute of Development Studies, several of whom were also involved in the writing of the I.L.O. report. See also their recent reader on the subject, *Third World Employment*, ed. Jolly, de Kadt, Singer and Wilson.

14. See H. Chenery and A. Strout, 'Foreign Assistance and Economic Development', *American Economic Review* (Sep 1966). These authors developed the two-gap model terminology. For a critique of the two-gap model see H. Burton, 'Two Gap Approach to Aid and Development', and H. Chenery, 'Comment and Reply', *American Economic Review* (June 1969).

15. Cf. 'Effects of Income Re-distribution on Economic Growth Constraints: Evidence from the Republic of Korea', *Economic Bulletin of Asia and the Far East*, vol. XXIII, no. 1 (June 1972).

16. See R. Dumont, with Marcel Mazoyer, *Socialisms and Development* (London: Andre Deutsch, 1973) especially chs 2 and 4.

CHAPTER EIGHT

1. K. Mannheim, *Ideology and Utopia* (London: Kegan Paul, French, Trubner, 1946) especially ch. II. The German concepts, however, are quoted from the German edition (1930) p. 47.

2. Ibid. especially pp. 175 ff.

3. Ibid. p. 177.

4. K. Mannheim, *Man and Society in an Age of Reconstruction*, 7th edn (London: Routledge & Kegan Paul, 1954) p. 178.

5. Ibid. pp. 180–1.

6. Quoted in Magdoff, *The Age of Imperialism*, p. 126.

7. These figures are based on the latest available directory of the 500 largest industrial corporations, as annually compiled for *Fortune* (May 1974).

8. D. L. Meadows *et al.*, *The Limits to Growth*, a Report for the Club of Rome Project on the Predicament of Mankind (New York: Universe Books, 1972).

9. See, for example, *Only One Earth*, the report of the Stockholm Conference (London: Earth Island, 1972); B. Ward and R. Dubois, also entitled *Only One Earth* (London: Andre Deutsch, 1972); and *A Blueprint for Survival*, an *Ecologist Magazine*'s Publication (London, 1972). Earlier work, signalling the same problems, came from Paul Ehrlich, *The Population Bomb* (London: Pan Books, 1971) first published in the United States in 1968. The latest in the series of gloom and doom predictors, at least in English, is the English version of

R. Dumont's *L'Utopie ou la Mort*, *Utopia or Else* (London: Andre Deutsch, 1974).

10. Raymond Aron, quoted in Dumont, *Utopia or Else*, p. 14.

11. The American Council for Environmental Quality speaks of between 4 and 5 billion dollars annually to be spent by industry on new anti-pollution measures. See Ward and Dubois, *Only One Earth*, p. 247. Since this estimate is about the equivalent of the total G.D.P. of most developing nations, the point made hardly needs pressing any further.

12. Dumont, *Utopia or Else.*

13. Mannheim, *Man and Society in an Age of Reconstruction*, p. 152.

14. Cf. a report in *Newsweek* (10 Feb 1975) by Varindra Tarzie Vittachi.

Author Index

Subject Index